Mitigating Process Variability and Soft Errors at Circuit-Level for FinFETs

Alexandra Zimpeck • Cristina Meinhardt
Laurent Artola • Ricardo Reis

Mitigating Process Variability and Soft Errors at Circuit-Level for FinFETs

 Springer

Alexandra Zimpeck (iD)
Universidade Católica de Pelotas
Pelotas
Rio Grande do Sul, Brazil

Laurent Artola (iD)
ONERA/DPHY University of Toulouse
Toulouse, France

Cristina Meinhardt (iD)
Universidade Federal de Santa Catarina
Florianópolis
Santa Catarina, Brazil

Ricardo Reis (iD)
Inst de Informatica, Bloco IV, CP15064
Univ Federal do Rio Grande do Sul
Porto Alegre
Rio Grande do Sul, Brazil

ISBN 978-3-030-68370-2   ISBN 978-3-030-68368-9   (eBook)
https://doi.org/10.1007/978-3-030-68368-9

This Springer imprint is published by the registered company Springer Nature Switzerland AG
The registered company address is: Gewerbestrasse 11, 6330 Cham, Switzerland

To my parents, Alexandre and Dulcinara, and my brother, Lucas, who always gave me unconditional love and all the support to follow my dreams.

"Alexandra Lackmann Zimpeck"

To my beloved parents, Armando Hugo and Laci, my dear sisters, Luciene and Simone, and my devoted husband, Odorico, for all the support and love.

"Cristina Meinhardt"

To the love of my life, Florie, and my two great children, Mathis and Tom, my devoted parents, Patricia and Jean, and my sister, Maïté, for all their support and love.

"Laurent Artola"

To my wife, Lucia, to my daughter, Mariana, to my sons, Guilherme and Eduardo, to my grandson, Noah, to my parents, Constantino and Maria de Lourdes, and to all students who are and were part of my research team.

"Ricardo Augusto da Luz Reis"

Preface

Process variability mitigation and radiation hardness are relevant reliability requirements as chip manufacturing advances into the nanometer regime. The unexpected behavior due to process variations can stimulate circuit degradation, making it inappropriate for their initial purpose. Moreover, a single energetic particle coming from terrestrial or space radiation can affect the proper operation of electronic devices, causing temporary data loss, among other effects. As the nanometer technologies increase the integration factor, a single particle hit can diffuse the charge to the adjacent nodes making the circuit even more prone to radiation effects. These facts can lead to financial and also human life losses, depending on the target application. These consequences emphasize the relevance of creating new design guidelines to deal with the reliability challenges imposed by current technologies.

In this scenario, this book evaluates the influence of process variations (e.g., work-function fluctuations) and radiation-induced soft errors in a set of logic cells using FinFET technology, considering the 7-nm technological node as a case study, which is explored by the renowned semiconductor industries currently. Moreover, this book adopts a radiation event generator tool (MUSCA SEP3) for accurate soft error estimation, which deals both with layout features and electrical properties of devices. Authors also explore four circuit-level techniques (e.g., transistor reordering, decoupling cells, Schmitt Trigger, and sleep transistor) as alternatives to attenuate the unwanted effects in FinFET logic cells. This book also evaluates the mitigation tendency when different levels of process variation, transistor sizing, and radiation particle characteristics are applied in the design. An overall comparison of all methods addressed by this book is provided, allowing to trace a trade-off between the reliability gains and the design penalties of each approach regarding the area, performance, power consumption, single event transient (SET) pulse width, and SET cross-section.

Pelotas, Brazil   Alexandra Zimpeck

Florianópolis, Brazil   Cristina Meinhardt

Toulouse, France   Laurent Artola

Porto Alegre, Brazil   Ricardo Reis

November 2020

Contents

Acronyms

3T	3-Terminal transistor
4T	4-Terminal transistor
AOI	AND–OR inverter
ASAP7	7-nm Predictive process design kit
BEOL	Back-end-of-line layer
BOX	Buried oxide
BTI	Bias-temperature instability
CAD	Computer-aided design
CMOS	Complementary metal-oxide-semiconductor
CMP	Chemical–mechanical planarization
DD	Displacement damage
DIBL	Drain-induced barrier lowering
DRC	Design rule check
DRM	Design rule manual
EDA	Electronic design automation
ELT	Enclosed layout transistor
EM	Electromigration
EUV	Extreme ultraviolet lithography
FA	Full adder
FEOL	Front-end-of-line layer
FER	Fin edge roughness
FinFET	Fin-shaped field-effect transistor
FO4	Fan-out 4
GAA	Gate-all-around
GDS	Geometric data stream
GER	Gate edge roughness
GSS	Gold standard simulation
HBD	Hardening by design
HCI	Hot carrier injection
HKMG	High-K metal gate
HP	High performance

IC	Integrated circuit
IG	Independent-gate
IHP	Innovations for high performance
ITRS	International Technology Roadmap for Semiconductors
LE	Logical effort
LELE	Litho-etch litho-etch
LEO	Low Earth orbit
LER	Line edge roughness
LET	Linear energy transfer
LIG	Local-interconnect gate
LISD	Local-interconnect source–drain
LP	Low power
LSTP	Low stand-by power
LVS	Layout versus schematic
MBU	Multiple-bit upset
MC	Monte Carlo
MGG	Metal gate granularity
MINLP	Mixed integer non-linear program
MOL	Middle-of-line layer
MOSFET	Metal-oxide-semiconductor field-effect transistor
MP	Multiple patterning
MTTF	Mean time to failure
MUSCA SEP3	Multi-Scales Single Event Phenomena Predictive Platform
NBTI	Negative bias-temperature instability
NFET	N-channel field-effect transistor
NIEL	Non-ionizing energy loss
OAI	OR–AND inverter
PDK	Process design kit
PDP	Power-delay-product
PFET	P-channel field-effect transistor
PTM	Predictive technology model
PTM-MG	Predictive technology model for multigate transistors
PVT	Process, voltage, and temperature
RC	Resistances and capacitances
RDF	Random dopant fluctuation
SADP	Self-aligned double-patterning
SAQP	Self-aligned quadruple-patterning
SCE	Short channel effect
SDT	Source–drain trench
SE	Soft error
SEB	Single event burnout
SEE	Single event effect
SEGR	Single event gate rupture
SEL	Single event latchup
SER	Soft error rate

SET	Single event transient
SEU	Single event upset
SG	Shorted-gate FinFET
SHE	Single hard error
SOI	Silicon on insulator
SPICE	Simulation Program with Integrated Circuit Emphasis
SRAM	Static random-access memory
STI	Shallow trench isolation
TCAD	Technology computer-aided design
TID	Total ionizing dose
TSMC	Taiwan Semiconductor Manufacturing Company
VCO	Voltage-controlled oscillator
VLSI	Very Large-Scale Integration
VTC	Voltage transfer curve
WF	Work-function
WFF	Work-function fluctuation
ZTC	Zero temperature coefficient

Chapter 1
Introduction

The technology scaling increases the transistor count in the same chip, satisfying the demand for higher density, lower cost, more functionalities, superior clock frequency, and reduced power consumption [2]. Nevertheless, it is harder to maintain the exponential growth rate in each new technology node, incurring higher design efforts and longer time to market. The high integration factor and the technology evolution brought new challenges for Very Large-Scale Integration (VLSI) designs.

Novel materials and new device architectures had to be implemented in the design of integrated circuits (ICs) to ensure the technology scaling sub-22nm [12]. The 3D structure and the lightly doped channel of fin-shaped field effect transistor (FinFET) devices imply a significant reduction of leakage currents, superior immunity to the short channel effects (SCEs), an increase of carrier mobility, and a reduction of random dopant fluctuations (RDFs) [1, 11]. These characteristics enhanced the channel's electrostatic control, one of the main challenges faced during the planar scaling. In this way, the adoption of FinFET devices brought several benefits for the semiconductor industry, maintaining the pace of advancement predicted by Moore's law. Figure 1.1 shows the technology scaling progress alongside the years, where metal-oxide-semiconductor field effect transistor (MOSFET) devices were widely used until 2011, being replaced by FinFET nodes shortly hereinafter.

The advanced technologies raise essential topics related to the reliability of electronic systems. At the sub-22nm nodes, the small geometric patterns increase the potential sources of process variability [6], and the higher density allows that a single energetic particle affects multiple adjacent nodes [4, 9]. Moreover, the lower supply voltages increase the sensitivity to the external noise. All these factors can compromise entire blocks of logic cells because they can modify the transistor structure and, consequently, the electrical properties, decreasing the IC robustness.

The process variability represents a random deviation from the typical design specifications that stimulates the circuit degradation, abnormal power consumption, and performance divergence [14]. FinFET technologies are more prone to process

© The Author(s), under exclusive license to Springer Nature Switzerland AG 2021
A. Zimpeck et al., *Mitigating Process Variability and Soft Errors at Circuit-Level for FinFETs*, https://doi.org/10.1007/978-3-030-68368-9_1

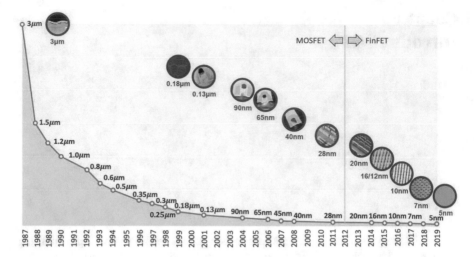

Fig. 1.1 Technology advancement alongside the years [15]

variations due to two main reasons: the wavelength used in the lithography step and the adoption of high-k dielectrics for improving the gate control on the channel region [8]. In the first case, as the wavelength has not kept pace with the technology scaling, the transfer of small geometric standards to the substrate surface results in a deviation in the device structure after the manufacturing process. On the other hand, the use of the metal gate modifies the grain orientations, generating different work-functions (WFs) aligned randomly, which implies higher process deviations [7].

The soft error (SE) arises from the interaction of energetic particles with the silicon coming from space and terrestrial radiations. In general, lower technology nodes decrease the minimum charge required to induce a single event transient (SET) pulse. It happens due to reduced nodal capacitance, low supply voltages, and higher frequency operations, increasing the probability of a memory element latching a SET generated in the combinational logic [3].

However, the FinFET disruptive nature is favorable to reduce the SE susceptibility. The connection between the transistor channel and the substrate occurs through a narrow region surrounded by isolation oxides, decreasing the volume of silicon exposed to the charge collection mechanism and modifying the sensitive areas. Consequently, this fact bounds the increase of soft errors in FinFET technologies [13]. Although the FinFET devices present attractive properties to control the radiation-induced soft errors, other reliability challenges, such as the process variability, can modify the linear energy transfer (LET) threshold to induce a soft error.

In this context, process variability mitigation and radiation hardness are relevant reliability requirements as chip manufacturing advances more in-depth into the nanometer regime. These challenges require a detailed and accurate evaluation considering several test scenarios and estimating the real impact on FinFET circuits from a design standpoint. Furthermore, the investigation and application of mitigation approaches are essential to ensure more reliable circuits.

1.1 Book Motivation

Integrated circuits under process variations can fail to meet some performance or power consumption criteria, leading to parametric yield loss and demanding several redesign steps. Radiation-induced soft errors can provoke temporary data loss, inducing critical system behavior failures even at the ground level. Transient faults can also result in human life losses depending on the target application. These consequences emphasize the importance of creating new design guidelines to deal with the challenges imposed by sub-22nm technologies. In the literature, several works explore the impact of process variability and soft errors in FinFET devices, but few research proposed solutions to attenuate the effects caused by them.

Both industry and academy pursue novel materials, structures, doping profiles, and design strategies to be implemented in the mainstream circuit to mitigate the impact of reliability challenges and continue with technology scaling using FinFET nodes. The debate on which of them is the best way is still open, but potential alternatives need to be pointed out and well investigated. Techniques related to manufacturing adjustments and hardware replication have been extensively explored. Although very efficient, manufacturing changes have an expensive cost with higher complexity involved, besides the hardware redundancy introduces large overheads.

However, much less understanding has been gained, focusing on physical design alternatives for FinFET technologies. Circuit-level approaches, which modify the classical design of standard cells, can be interesting to achieve more robust solutions with fewer implementation costs and penalties. The physical design adjustments are related to the insertion of some components, transistor reordering or sizing, modifications in a pull-up or pull-down network, scaling up the supply voltage, different gate topologies that implement the same function, etc.

1.2 Goals

Given this context, the overall purpose of this book is to promote four circuit-level approaches to reduce the consequences caused by process variations and radiation-induced soft errors in FinFET-based circuits, highlighting all the pros and cons of adopting it. This work is divided into the following steps:

- evaluation of FinFET logic cells at standard conditions, under process variations (e.g., work-function fluctuations) and radiation effects (e.g., soft errors);

- application of four circuit-level mitigation techniques at the layout level to obtain more reliable FinFET logic cells;
- verification of the mitigation tendency of process variability when different levels of variations and transistor sizing are applied in the most reliable design (e.g., using circuit-level methods);
- assessment of the mitigation tendency of soft error susceptibility when different radiation particle characteristics, such as LET, are added in the reliable design with different input vectors and core voltage;
- achievement of a trade-off between the gains and penalties of each approach regarding the area, performance, power consumption, SET pulse width, and SET cross section;
- provision of an overall comparison between all techniques applied in this work and those available in the literature.

The four circuit-level techniques explored in this book are called (1) transistor reordering, (2) sleep transistor, (3) decoupling cells, and (4) Schmitt Trigger. The first two techniques are based on changes in the pull-up or pull-down networks, and the other two consist of connecting some components in the circuit output. In summary, the focus of the transistor reordering method is the optimization of the transistor arrangements. In contrast, the insertion of a sleep transistor is targeted at improving the overall power. Decoupling cells and Schmitt Triggers can be considered capacitive methods. They are commonly used to minimize output degradations. The potential of methods is assessed by comparing the well-known predefined metrics of the standard design with the design adopting each technique.

As differential factors, this book provides all results considering a predictive 7-nm FinFET process design kit (PDK) [5], which is the same technology node explored by the renowned semiconductor industries currently. Moreover, for accurate soft error estimation, this book also adopts a Multi-Scales Single Event Phenomena Predictive Platform (MUSCA SEP3), which deals with layout features coupled with radiation-induced currents modeled by the tool and electrical properties of devices [10]. After reading this book, the readers will be able to:

- comprehend the influence of process variability (e.g., work-function fluctuations) and radiation-induced soft errors in FinFET logic cells;
- learn how it is possible to improve the robustness of FinFET integrated circuits without focusing on manufacturing adjustments;
- understand the benefits and downsides of using circuit-level approaches such as transistor reordering, decoupling cells, Schmitt Trigger, and sleep transistor for mitigating the impact of process variability and soft errors;
- verify the capability of the promoted techniques when different test scenarios were considered such as distinct levels of process variations, transistor sizing, and different radiation features;
- identify the best technique considering the target application and design requirements like area constraints, power/delay limitations, or radiation hardness.

1.3 Outline

This book is organized into nine chapters. This chapter focuses on introducing the reliability issues in modern circuit design and presenting the book's goals. The following paragraphs give a short book summary chapter by chapter.

Chapter 2—FinFET Technology: This chapter introduces the basic concepts related to FinFET nodes. The structure and properties of FinFET devices, the advancement in the semiconductor industry, and layout considerations are also explored. Moreover, this chapter presents and discusses some predictive models and PDKs available for academic use.

Chapter 3—Reliability Challenges on FinFETs: This chapter discusses several aspects of two reliability challenges in FinFET electronic devices: process variability and radiation-induced soft errors. The most relevant state-of-the-art works are presented, focusing on evaluating or attenuating nanometer technologies' reliability challenges.

Chapter 4—Circuit-Level Mitigation Approaches: This chapter explains the four circuit-level mitigation techniques promoted in this work: transistor reordering, decoupling cells, Schmitt Trigger, and sleep transistor. The design implementation aspects of each of them are also discussed in this chapter.

Chapter 5—Evaluation Methodology: This chapter provides the complete design flow adopted in this work, highlighting the main differences between a traditional design and a design focused on evaluating the impact of process variations and radiation susceptibility.

Chapter 6—Process Variability Mitigation: This chapter demonstrates the potential of each circuit-level technique regarding process variability attenuation comparing with the typical behavior of FinFET logic cells. The chapter also presents the mitigation tendency when different levels of process variations and transistor sizing were adopted.

Chapter 7—Soft Error Mitigation: This chapter evaluates the potential of each circuit-level technique to reduce the soft error susceptibility of FinFET logic cells. A comparison between the SET cross section and SET pulse width is showed considering different LET values, supply voltages, and input vectors.

Chapter 8—General Trade-Offs: This chapter presents the drawbacks involved with the adoption of circuit-level mitigation techniques regarding the area, power consumption, and propagation delays. An overall comparison is provided, allowing to trace a trade-off between the gains and penalties of using each approach.

Chapter 9—Final Remarks: This chapter summarizes a set of considerations that reinforce the importance of this book's contributions. Some possibilities of relevant open research are also indicated in this chapter.

References

1. Agostinelli, M., Alioto, M., Esseni, D., Selmi, L.: Design and evaluation of mixed 3t-4t FinFET stacks for leakage reduction. In: Svensson, L., Monteiro, J. (eds.) Integrated Circuit and System Design. Power and Timing Modeling, Optimization and Simulation, pp. 31–41. Springer, Berlin (2009)
2. Anghel, L., Rebaudengo, M., Reorda, M.S., Violante, M.: Multi-level Fault Effects Evaluation, pp. 69–88. Springer, Dordrecht (2007)
3. Baumann, R.C.: Radiation-induced soft errors in advanced semiconductor technologies. IEEE Trans. Device Mat. Reliab. **5**(3), 305–316 (2005)
4. Bhuva, B.L., Tam, N., Massengill, L.W., Ball, D., Chatterjee, I., McCurdy, M., Alles, M.L.: Multi-cell soft errors at advanced technology nodes. IEEE Trans. Nuclear Sci. **62**(6), 2585–2591 (2015)
5. Clark, L.T., Vashishtha, V., Shifren, L., Gujja, A., Sinha, S., Cline, B., Ramamurthy, C., Yeric, G.: ASAP7: A 7-nm FinFET predictive process design kit. Microelectron. J. **53**, 105–115 (2016)
6. Collins, L.: FinFET variability issues challenge advantages of new process (2014). https://www.techdesignforums.com/blog/2014/04/16/FinFET-variability-challenges-advantages/
7. Dadgour, H.F., Endo, K., De, V.K., Banerjee, K.: Grain-orientation induced work function variation in nanoscale metal-gate transistors–part ii: Implications for process, device, and circuit design. IEEE Trans. Electron Devices **57**(10), 2515–2525 (2010)
8. Deshmukh, R., Khanzode, A., Kakde, S., Shah, N.: Compairing FinFETs: SOI vs bulk: Process variability, process cost, and device performance. In: 2015 International Conference on Computer, Communication and Control (IC4), pp. 1–4 (2015)
9. Endo, K., Matsukawa, T., Ishikawa, Y., Liu, Y.X., O'uchi, S., Sakamoto, K., Tsukada, J., Yamauchi, H., Masahara, M.: Variation analysis of tin FinFETs. In: 2009 International Semiconductor Device Research Symposium, pp. 1–2 (2009)
10. Hubert, G., Duzellier, S., Inguimbert, C., Boatella-Polo, C., Bezerra, F., Ecoffet, R.: Operational SER calculations on the SAC-C orbit using the multi-scales single event phenomena predictive platform (MUSCA SEP3). IEEE Trans. Nuclear Sci. **56**(6), 3032–3042 (2009)
11. King, T.-J.: FinFETs for nanoscale CMOS digital integrated circuits. In: ICCAD-2005. IEEE/ACM International Conference on Computer-Aided Design, 2005., pp. 207–210 (2005)
12. Pradhan, K.P., Sahu, P.K., Ranjan, R.: Investigation on asymmetric dual-k spacer (ads) trigate wavy FinFET: A novel device. In: 2016 3rd International Conference on Devices, Circuits and Systems (ICDCS), pp. 137–140 (2016)
13. Seifert, N., Jahinuzzaman, S., Velamala, J., Ascazubi, R., Patel, N., Gill, B., Basile, J., Hicks, J.: Soft error rate improvements in 14-nm technology featuring second-generation 3d tri-gate transistors. IEEE Trans. Nuclear Sci. **62**(6), 2570–2577 (2015)
14. Tassis, D., Messaris, I., Fasarakis, N., Nikolaidis, S., Ghibaudo, G., Dimitriadis, C.: Variability analysis – prediction method for nanoscale triple gate FinFETs. In: 2014 29th International Conference on Microelectronics Proceedings - MIEL 2014, pp. 99–102 (2014)
15. TSMC: 5nm Technology (2019). https://www.tsmc.com/english/dedicatedFoundry/technology/logic

Chapter 2
FinFET Technology

For several decades, the planar complementary metal-oxide-semiconductor (CMOS) technology was the main core of integrated circuits, but MOSFET devices reached the physical limits [24, 61]. The gate control over the channel region suffers changes as the gate electrode is reduced. MOSFET devices required high channel doping to control the SCE, reflecting in the mobility degradation and a significant increase of leakage currents, affecting the transistor performance directly [34]. Novel materials and new device architectures need to be adopted for ensuring the advancement of microelectronics in sub-22nm nodes [53].

Multigate devices gained prominence for presenting better SCE control, reduced leakage currents, high driving capability, and a better yield [45]. Hence, the International Technology Roadmap for Semiconductors (ITRS) pointed them as the most attractive choice to overcome obstacles and keep scaling [34, 37]. There are a variety of multigate devices available in the literature such as the π-gate [50], Ω-gate [79], gate-all-around (GAA) [68], FlexFET [77], and Trigate [11, 35]. However, the FinFET was predominantly adopted mainly due to the similarity of the manufacturing process with conventional planar technologies.

This chapter will address the main aspects of FinFET devices once they are still extensively studied in academia and used in industrial processes. First, we will explore the structure, properties, and expected behavior of FinFETs. Then, we will explain the advancement of the semiconductor industry and the next tendencies involving FinFET nodes. Finally, we will evaluate the design rules for layout generation and the predictive models available in the literature for academic purposes.

© The Author(s), under exclusive license to Springer Nature Switzerland AG 2021
A. Zimpeck et al., *Mitigating Process Variability and Soft Errors at Circuit-Level for FinFETs*, https://doi.org/10.1007/978-3-030-68368-9_2

2.1   Structure and Properties

The idea of multiple-gate transistors to reduce the SCE was first announced by Sekigawa and Hayashi in 1984 [65]. A novel multigate device called DELTA was proposed in 1989 with a vertical ultra-thin structure. This formation provides better channel controllability, higher transconductance, and minimized subthreshold swing [30].

Hisamoto investigated the first FinFET with silicon-on-insulator (SOI) substrate [31]. The author observed that the devices need to be self-aligned to one another and with the source/drain terminals for reducing the parasitic resistances. The first experimental evidence using FinFET devices was a four-stage inverter chain [55] and static random-access memory (SRAM) cell [49]. Since then, FinFET devices have been widely explored in the last decades.

Figure 2.1 shows a basic comparison between the planar and FinFET structures. A FinFET device consists of a vertical silicon fin to form the channel region and connect the source and drain regions at each end. The vertical fin wraps the gate region, and a MOS channel is formed at the two sidewalls plus the top side of the fin. More information about the main differences between MOSFET and FinFET structures can be encountered in [23, 24, 32, 39, 48, 70, 71].

The fin-like geometry implies no free charge carriers available, allowing the suppression of SCE [37]. Unlike MOSFET devices, FinFET channels have lower doping profiles ensuring better mobility of the carriers and better performance. A detailed explanation about the improvements of SCE in FinFET devices over traditional MOSFETs is discussed in [27, 31, 62, 74, 80].

The adoption of FinFET devices in the manufacturing process also improves leakage power significantly. In Fig. 2.2, it is possible to observe a little increase in leakage power of sub-10nm FinFET nodes. However, the dynamic power

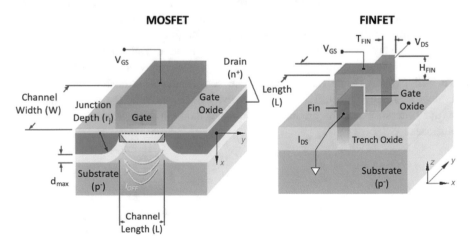

Fig. 2.1 Structural differences between MOSFET and FinFET devices [20]

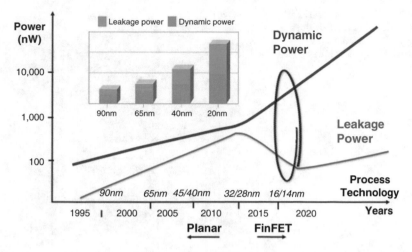

Fig. 2.2 Planar and FinFET power consumption trends [44, 57]

dissipation keeps growing as more transistors being packed together, ensuring the best performance for the circuits [57]. In general, FinFET devices offer interesting power-delay trade-offs with advantageous characteristics for both low-power and high-performance applications [45]. The aspects of FinFET circuit design and how it impacts the performance, power consumption, and area can be better explored in [8, 16, 27, 37, 45, 52, 73, 84].

FinFETs include additional properties to describe the fin configurations besides the length and width, typical measurements in planar nodes. The key geometric parameters for a FinFET are the gate length (L_G), the fin height (H_{FIN}), the fin thickness ($T_{SI}/T_{FIN}/W_{FIN}$), and the oxide thickness (T_{OX}), according to Fig. 2.3a and b. Fin engineering (balancing the fin height, the fin thickness, the oxide thickness, and the channel length) is essential to minimize the leakage current, I_{OFF}, and maximize the on-state current, I_{ON} [73].

Higher fin height values can result in structural instability that impacts the manufacturing process and the SCE control. On the other hand, smaller fin height offers more flexibility, but this leads to more silicon area due to the need for multiple fins [8]. The effective channel length (L_{EFF}) and width (W_{EFF}) of a single FinFET double-gate transistor are given by Eqs. 2.1, and 2.2, respectively. The L_{EXT} corresponds to the full extension of the source and drain regions, and the W_{MIN} of a FinFET device is approximately equal to $2 \times H_{FIN}$.

$$L_{EFF} = L_G + 2 \times L_{EXT} \tag{2.1}$$

$$W_{EFF} = T_{SI} + W_{MIN} \tag{2.2}$$

In planar technologies, the transistor channel width can receive arbitrary values as long as it obeys the design constraints. For FinFETs, the channel width has a quantization characteristic using a discrete sizing [48]. For scaling the effective

Fig. 2.3 The key geometric parameters of FinFET devices considering (**a**) a 3D view and (**b**) the top view [18]

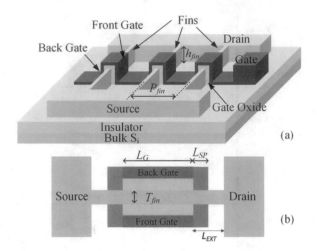

channel width ($W > W_{MIN}$), it is necessary to increase the number of fins connected in parallel, which are sharing a common lateral diffusion, as shown in Fig. 2.3a. The width of a FinFET with multiple fins is given by Eq. 2.3, where N_{FIN} is the number of fins. The FinFET electrical characterization such as threshold voltage analysis and $I_D \times V_G$ curves can be seen in [10, 12, 28, 60, 75].

$$W = N_{FIN} \times W_{MIN} \tag{2.3}$$

FinFET devices can be designed using different configurations for the gate terminals and also for the substrate. About the gates, FinFETs can be classified as shorted-gate (SG) or independent-gate (IG) [45]. In the first case, the back-gate and the front-gate are tied together, physically shorted. The SG structure is very similar to MOSFET devices, with three terminals (3T) controlling the transistor operation. In this mode, both gates provide maximum gate drive, making an excellent electrostatic control of the channel. However, the off-current is more elevated because it cannot regulate the device's threshold voltage (V_{TH}) electrically.

The IG FinFET modifies the concept previously presented. The top part of the gate is etched out, generating two independent gates to control the channel. This format allows each gate to have a different input signal, generating a four-terminal (4T) transistor [8]. Typically, back-gates are used to control the threshold voltage of front-gates to obtain even smaller leakage currents. Moreover, this configuration offers the designers more flexibility to create low-power circuits [59]. The structural differences between SG and IG FinFET devices are shown in Fig. 2.4. More information about the benefits, drawbacks, and physical issues of SG and IG FinFET devices can be encountered in [9, 72, 76]. Novel circuit styles using SG and IG FinFETs for low-power design and leakage current reduction were carefully studied in [1, 47, 59, 83].

FinFET devices can be fabricated on conventional bulk or in SOI substrates, as illustrated in Fig. 2.5. In bulk FinFETs, all fins share a common silicon substrate,

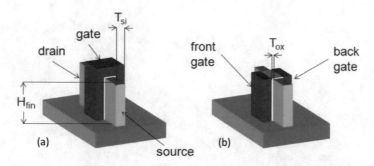

Fig. 2.4 FinFET with (**a**) shorted-gate and (**b**) independent-gate configurations [67]

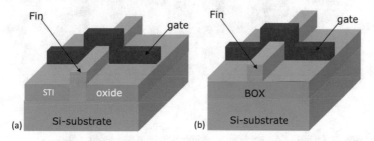

Fig. 2.5 Comparison between (**a**) bulk and (**b**) SOI substrates [17]

and the shallow trench isolation (STI) oxide provides the insulation between the adjacent fins [8]. On the other hand, the SOI FinFET has a thin layer of oxide (SiO_2), and the devices are manufactured on the top of the buried oxide (BOX). Bulk FinFETs present a set of benefits: less expansive wafers, low defect density, less back-gate bias effect, better immunity to heat transfer problems, lower negative bias-temperature instability (NBTI) stress, and similarity with the planar fabrication processes [6, 51]. The disadvantage is the fin formation that happens by a timed etch process, more prone to process variations. Moreover, this kind of substrate requires stronger doping that impacts the leakage currents, causing mobility degradation. In SOI FinFETs, the fin structure is formed through a natural process that stops when it reaches the buried oxide layer, reducing process variations [56]. Furthermore, SOI substrates minimize parasitic capacitances and improve the current drive, circuit speed, and power consumption [16].

2.2 Advancement of Semiconductor Industry

In 1965, Gordon E. Moore proposed the famous Moore's law. He predicted that the transistor count on a chip doubles every two years with the same cost and improvements in the transistor performance [46]. Until now, the semiconductor

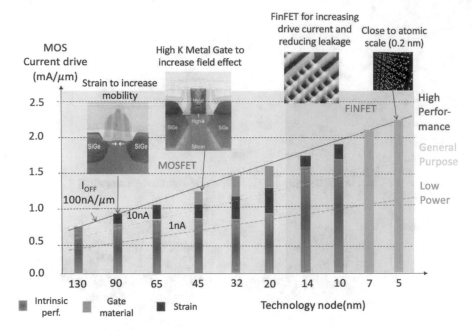

Fig. 2.6 Advancements of the nanometer regime in the last few years [66]

industry adopts this prediction as a guide and, from it, sets targets for research and development. Figure 2.6 illustrates the progress of the nanometer regime and some techniques and materials used to keep scaling over the last few years. The high price of research, development, and manufacturing equipment led to a reduction in the number of semiconductor industries investing in advanced technological nodes. As an example, the GlobalFoundries recently decided to stop the design of chips at the 7-nm FinFET technology. Nowadays, the race for more compact and technological chips happens mainly among the Intel, Samsung, and Taiwan Semiconductor Manufacturing Company (TSMC) [29].

The Intel Corporation was the first company to produce microprocessors using a FinFET technology [5]. Both mobile and desktop processors adopted the 22-nm node and became available in the market in 2012. Ivy Bridge is the name used to represent all the first-generation Intel Core processors based on FinFET technology. Other semiconductor industries have introduced FinFET circuits in the market since 2014.

TSMC started the volume production of 16-nm FinFET node in 2014, obtaining circuits 50% faster, with 60% less power consumption. Shortly after, Intel introduced the 14-nm FinFET node. Compared with the first node, the fins of transistors are taller, thinner, and more closely spaced to improve the density and decrease the capacitances [64]. In the same year, Exynos7 was introduced in the market as the first mobile processor developed by Samsung on the 14-nm FinFET technology, offering about 20% more performance and 35% power reduction than existing nodes.

In 2016, Samsung, Intel, and TSMC announced the 10-nm FinFET node in the consumer market. Samsung devices improved the area efficiency by 30%, increased the performance by 27%, and reduced the power consumption by 40% [63]. Intel Corporation presented circuits with 25% better performance and 45% lower power consumption [33]. For TSMC, the new node improved 2× logic density along with 15% faster speed and 35% less power consumption. In 2017, the second generation of the 10-nm FinFET nodes with even more benefits was made available by Intel and Samsung.

The mass production of 7-nm devices began in 2018. The TSMC launched 7-nm devices for mobile and high-performance computing applications with improvements of 1.6× logic density, 20% of higher speed, and 40% of power reduction. In 2019, Samsung announced the beginning of wafer production using the 7-nm process using multiple patterning for selected layers. This technology enables a 40% area reduction along with 50% lower power and 20% higher performance compared to the 10-nm process.

Recently, Intel executives confirmed that the company would not launch processors with 7-nm lithography until the end of 2021. For technology nodes below 10 nm, the most important metric is the transistor density. Thus, the 7-nm process offered by Samsung and TSMC is roughly equivalent to Intel's 10-nm process. Figure 2.7 reinforces this statement and illustrates the transistor density comparison between the three more relevant semiconductor industries.

In 2019, TSMC delivered the first complete design infrastructure for a 5-nm process technology. The volume production started in the first half of 2020. In the same way, Samsung informed that the 5-nm FinFET technology is ready for customer's samples. This node reduces the mask layers presenting 25% area efficiency and 20% lower power or 10% higher performance. However, the scaling benefits are questionable for these 5-nm FinFET nodes. They are considered half-nodes because they do not provide the double density and significant improvements in power and performance metrics.

According to the consumer's point of view, they wish for a product that offers better performance for the lowest price with a reliable roadmap for future generations. For chips to get denser, the chipmakers have two options: try to extend FinFETs to 3 nm or migrate to gate-all-around devices. The thinner geometric patterns imposed on each new FinFET node increase the probability of phenomena like line edge roughness (LER) and metal gate granularity (MGG) that modifies the ideal shape and features of the transistor channel preventing the transistors from working as desired. The deviation in the structure due to the process variability impacts mainly the performance of a transistor. For this reason, to continue the scaling with FinFET devices, it is necessary to mitigate the effects of process variability or find a performance booster able to control the weak electrostatic of the channel.

On the other hand, the adoption of GAA devices implies technical and cost challenges. The device manufacturing process requires an extra step that involves extreme ultraviolet (EUV) lithography with a high level of complexity and multiple reliability issues. However, the implementation demands huge funding, putting the

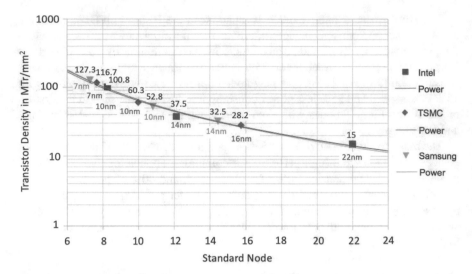

Fig. 2.7 Transistor density regarding the technology scaling in commercial FinFET technologies [29]

technology out of reach for many consumer applications. With all these drawbacks in mind, the introduction of another node or the switching to an alternative device may be delayed beyond the target date of 2021. It was recently announced that adverse to Samsung's 3-nm process node, which will use GAA transistors, TSMC will be sticking with FinFET devices relying on innovative features to achieve the full-node scaling [25].

2.3 Layout Design

In general, the essential difference in a FinFET manufacturing process is the existence of fins. However, from a physical point of view, the procedure of fin formation is not a trivial task and requires high aspect ratio etches and higher stress for mobility enhancements imposing challenges in the design [36]. Figure 2.8 highlights the differences between the layout of a transistor designed using MOSFET and FinFET technologies.

In FinFET layouts, the conducting channel between the source and drain terminals is formed through the fins with local interconnect layers. The area occupied by fins in a shorted-gate FinFET is given by Eq. 2.4, where P_{FIN} is the fin pitch, which corresponds to the distance between the middle section of two parallel fins. The disadvantage at the layout level for independent-gate devices is the area needed for placing the two separated contacts of the gates. Hereafter, some experimental results are presented, exploring the fundamental concepts of FinFETs at the layout level and methods to improve the layout density.

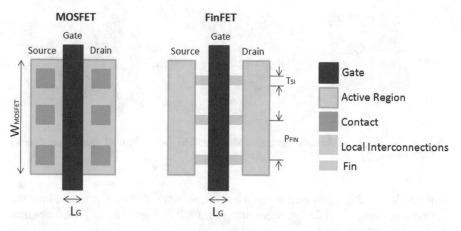

Fig. 2.8 Layout comparison of a MOSFET and an SG FinFET with three fins [19]

$$A_{FIN} = (N_{FIN} - 1) \times P_{FIN} \qquad (2.4)$$

According to Eq. 2.4, there are two ways to improve the FinFET transistor area: reducing the fin pitch or the number of fins. The fin pitch can be defined through the lithography-defined and spacer-defined methodologies [2, 4]. In the first case, the fin pitch minimum value is set by the adopted technology node. Otherwise, in the spacer-defined technique, the fin pitch can be halved due to an additional lithography step. The replacement of W_{MIN} by $2 \times H_{FIN}$ in Eq. 2.3 indicates that the number of fins can be reduced by increasing the fin height. However, the increase of H_{FIN} is restricted due to practical considerations. The acceptable ratio of H_{FIN}/T_{SI} in FinFETs should be around 2 [16].

Datta et al. presented an inverter layout in a IG model to verify the area penalty imposed by a 4T device considering a 45-nm technology node [21]. They represent the area of the cell concerning the minimum spacing requirement (λ) proposed by Nowak et al. [48]. The inverter area suffers an increase of 9.1% with IG structure due to the additional back-gate contact. A similar area penalty can also be observed in the NAND2 and NOR2 logic cells. However, in merged cells, the reduction in the number of transistors provoked an area contraction when compared to SG devices, mainly in the low-power circuit design. For ISCAS85 benchmark circuits, the IG devices obtained around 8.5% of area economy and 18% of power saving over the SG structure with no performance penalty.

Kumar and Kirubaraj investigated four FinFET logic design styles, such as shorted-gate, independent-gate, low power (LP), and hybrid (IG/LP) for a NAND2 cell [40]. Moreover, rectangular and fin-shaped diffusion approaches were considered for all FinFET modes presented above. Power dissipation and performance were analyzed for each design style considering different supply voltages. The results were compared with the NAND2 cell implemented using traditional planar technology. In general, they conclude that power dissipation is smaller in the

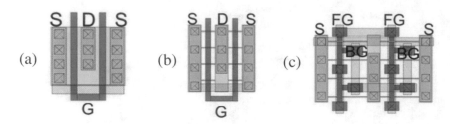

Fig. 2.9 Two-finger structure for (**a**) bulk CMOS devices, (**b**) 3T FinFET devices, and (**c**) 4T FinFET devices [3]

FinFETs logic design and even better when a fin-shaped diffusion is used. However, the average delay of all design styles using FinFET devices is more prominent than traditional planar technologies.

Alioto compared the lithography- and spacer-defined methods in 3T devices and the dependence of layout density with the fin height considering geometric constraints imposed by the standard cell library [2]. The FinFET layout density is better than planar CMOS, even for moderately tall fins. The lithography-defined (spacer-defined) FinFET cells exhibit an average area reduction around 0.95/0.58 = 1.64 (0.68/0.52 = 1.31) when increasing H_{FIN} from T_{SI} to $4T_{SI}$. Hence, fin height is considered a powerful knob to improve the layout density in FinFET cells.

Alioto compared the layout density of 3T, 4T, and mixed 3T–4T FinFET devices [3]. His results demonstrated that 3T and MT devices in the standard cell format have the same layout density as planar cells for low fin height values. The results were even better when a moderate fin height was applied. Instead, 4T devices have an unfavorable layout density due to the front- and back-gates' separated contacts. Hence, the fin pitch must be higher than the minimum value allowed by the technology. The multi-finger layout structure used to implement very wide transistors was also evaluated. The two-finger 3T transistor has a 15% lower area ratio compared with the traditional devices with $H_{FIN}/T_{SI} = 2$. Figure 2.9 shows the two-finger structure for planar CMOS, 3T, and 4T FinFET layouts.

Wimer proposed an algorithm for converting planar transistors in multigate devices, maximizing the number of fins in the target cell [78]. This algorithm is also applicable from a multigate source to a target layout for future technology nodes. He used a hard-IP technique to ensure the migration of a physical layout of an existing chip into a new target technology, with the same functionality and preserving all layout design rules. The place and route maintain the relative positions as in the source layout. The conversion flow has been successfully tested on full adders and sequential circuits.

McLellan presented a detailed comparison between the layout of an inverter designed in planar CMOS and FinFET technologies [42], as shown in Fig. 2.10. The transistors have a number of fins equal to five. The FinFET design is composed of rows of source/drain with gate strips orthogonally. Single gates usually violate the design rules of FinFET technologies. Thus, the FinFET inverter has three gates

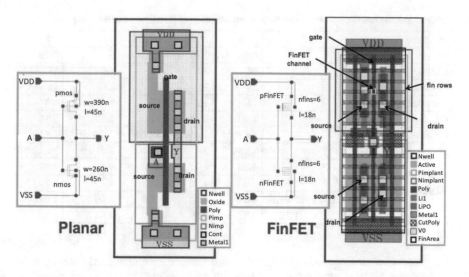

Fig. 2.10 Planar to FinFET layout differences [42]

(red) because every transistor must be finished with dummy gates on either side. It is impossible to cut off the diffusion by just ending the polygon as in the planar devices. The red hashed area in the middle of the FinFET inverter is the cut mask that separates the PFET and NFET transistors. Some geometric restrictions imposed by standard cell methodologies were presented in [2, 13, 38, 82].

Cui et al. built a standard cell library in the 7-nm FinFET technology, choosing an adequate number of fins for the complementary networks of all logic cells [19]. The standard cell library contains ten combinational logic cells with different drive strengths and three sequential cells activated on the positive edge. The λ-based layout design represents the characterization of each cell with the same height. The power density report was made, including a near- and super-threshold operation for different ISCAS benchmarks. Results show that the power density of the 7-nm node is at least 10–20 times larger than the same circuit in the 45-nm planar technology. Then, thermal management is necessary for FinFET nodes.

2.4 Predictive Models

With the continuous advancement of technology, predictive models are essential in circuit designs to identify design requirements and explore challenges and possible solutions [69]. The basic idea is to consider previous results to develop a representative model that can be used to predict future outcomes. Multigate transistors have been extensively evaluated through technology computer-aided design (TCAD) simulation tools providing high precision. However, the simulation time for VLSI circuits can be huge. Electrical models for Simulation Program with

Table 2.1 Geometric parameters and doping information of PTM-MG models [69]

Parameters			20 nm	16 nm	14 nm	10 nm	7 nm
Supply voltage (V)			0.90	0.85	0.80	0.75	0.70
Gate length (nm)			24	20	18	14	11
Fin height (nm)			28	26	23	21	18
Fin thickness (nm)			15	12	10	8	6.5
Oxide thickness (nm)			1.40	1.35	1.30	1.20	1.15
Channel doping (m^{-3})			5×10^{23}	1×10^{23}	5×10^{22}	2.5×10^{22}	1×10^{22}
Source–drain doping (m^{-3})			3×10^{26}	3×10^{26}	3×10^{26}	3×10^{26}	3×10^{26}
Work-function (eV)	HP	NFET	4.38	4.41	4.42	4.42	4.42
		PFET	4.80	4.76	4.75	4.75	4.74
	LSTP	NFET	4.56	4.58	4.60	4.60	4.61
		PFET	4.62	4.59	4.57	4.56	4.56

Integrated Circuit Emphasis (SPICE) require shorter computational time than 3D models, providing an alternative to aid circuit designers [43].

Some predictive models based on SPICE simulations were developed for FinFET technologies [82, 84]. However, these models do not represent an accurate transistor behavior needing some adjustments to be widely adopted. In 2012, a new generation of predictive technology model (PTM) for multigate transistors into sub-20nm technology nodes called PTM-MG was developed by Arizona State University (ASU) [54, 69]. These models provide high-performance (HP) and low stand-by power (LSTP) versions, where the main difference among them is the threshold voltage adopted. The flow to generate the predictive models was based on the TCAD model, where the main parameters were scaled considering the ITRS trends. The reference values of geometric parameters and doping information for a set of predictive nodes from PTM are shown in Table 2.1. The main disadvantage of PTM is not allowing the simulation of FinFETs in the IG structure. After, another model based on PTM-MG was proposed to fill this gap allowing the circuit simulation with different voltage in the front-gate and back-gate terminals [81].

In 2016, another predictive model called ASAP7 developed by ASU in partnership with ARM Ltd became available to estimate the trends of the 7-nm FinFET technology [15]. This model is considered more accurate than the previous one because it considers realistic conjectures from the semiconductor industry. The ASAP7 provides the typical–typical (TT), fast–fast (FF), and slow–slow (SS) process models for the designers. However, the FF and SS models affect the I_{ON}/I_{OFF} currents considerably. Table 2.2 shows a summary of the main geometric parameters and the doping characteristics using the ASAP7 model.

Both PTM and ASAP7 models allow the simulation of different FinFET structures (double-gate, triple-gate, or quadruple-gate) and the substrate mode (bulk or SOI). It is necessary to set the geomode and bulkmode parameters correctly in the technology file according to Table 2.3.

Table 2.2 Geometric parameters and doping information of ASAP7 models [15]

Parameters		TT	FF	SS
Supply voltage (V)			0.7	
Gate length (nm)			21	
Fin height (nm)		32	34	30
Fin thickness (nm)		6.5	7	6
Oxide thickness (nm)			2.1	
Channel doping (m^{-3})			1×10^{22}	
Source–drain doping (m^{-3})			2×10^{26}	
Work-function (eV)	NFET		4.37	
	PFET		4.81	

Table 2.3 Setup for different gate configurations and substrate [14]

	Geomode	Bulkmode
0	Double-gate	SOI
1	Triple-gate	Bulk
2	Quadruple-gate	–

The FinFET PDK, cell libraries, and design flow used by the semiconductor industries are not available for academic use. In this way, the North Carolina State University (NCSU) and the ASU in collaboration with ARM Ltd proposed free and predictive PDKs exploring the 15-nm and 7-nm nodes, respectively [7, 15]. Both PDKs are not tied to any specific foundry. This book focuses on the ASAP7 PDK because it allows design exploration at the 7-nm node. It is the current technology used in the manufacturing process of the largest semiconductor industries. Moreover, the developers considered realistic design conjectures regarding lithography steps and the current technology competencies of commercial nodes. Table 2.4 shows a summary of the widths and pitches beside the kind of lithography adopted for all layers from the ASAP7 PDK.

The lithography step uses ultraviolet light to transfer the geometric patterns to the thin wafers of silicon. FinFET technologies tried to adopt single exposure and EUV for all layers to provide cost-effective and straightforward designs. However, the wavelength has not kept pace with the device scaling, introducing challenges to print the small standards required when the technology nodes reach 14 nm and beyond [58]. Multiple patterning (MP) is the method used to overcome some lithography limitations and ensure enough resolution in the manufacturing process [41].

Different types of multiple patterning can be requested, depending on the circumstances. The implementation of MP can be through double or quadruple patterning. Double patterning involves two lithographic exposures and etches steps processed separately where the final shapes are combined at the end to create a single layer [26]. The litho-etch-litho-etch (LELE) and self-aligned double-patterning (SADP) are the most common ways to implement double patterning. The SADP has a significant advantage in the overlay tolerance if compared with LELE. The self-aligned quadruple patterning (SAQP) is a natural extension of SADP such that the only difference is one more step of spacer deposition [22]. Intel and TSMC already incorporated the SAQP methodology in their most recent technology nodes.

Table 2.4 Key layer
lithography assumptions,
widths, and pitches [15]

Layer	Lithography	Width/drawn (nm)	Pitch (nm)
Fin	SAQP	6.5/7	27
Active	EUV	54/16	108
Gate	SADP	21/20	54
SDT/LISD	EUV	25/24	54[a]
LIG	EUV	16/16	54
VIA0-VIA3	EUV	18/18	25[b]
M1-M3	EUV	18/18	36
M4-M5	SADP	24/24	48
VIA4-VIA5	LELE	24/24	34[b]
M6-M7	SADP	32/32	64
VIA6-VIA7	LELE	32/32	45[b]
M8-M9	SE	40/40	80
VIA8	SE	40/40	57[b]

[a] Horizontal only
[b] Corner-to-corner spacing as drawn

The manufacturing process of FinFET technologies is divided into three categories: front-end-of-line (FEOL), middle-of-line (MOL), and back-end-of-line (BEOL). The first group includes the production of wells and transistors, with the essential elements as the active region, fins, gate, and diffusions. The BEOL considers the contacts via layer and the metallic layers from metal 1 (M1) to the top metal (M9). This stage is responsible for short connections and overall cell routing. The connection between FEOL and BEOL steps happens in the MOL stage. For example, the source–drain trench (SDT) layer connects the active area to the local interconnect source–drain (LISD) layer, and the LISD joins the source and drain terminals of transistors. LISD is above SDT in the MOL stack. The local interconnect gate (LIG) is used for the contacts of the gate terminal. The purpose of V0 is to join the LIG and LISD to the BEOL layers.

Figure 2.11 shows the basic design rules of ASAP7 PDK. There is a set of details to be considered in the FinFET layout design. The fins need to be uniformly aligned, respecting the fixed fin pitch and the exact vertical fin width. Moreover, all fin layer polygons should have an equal length along the horizontal axis, and they cannot be bent. The FinFET fabrication process uses two dummy gates at each end of the cells, not allowing a single gate in the design. The gate layer with bends is also not supported. Each gate must have an exact fin pitch and a fixed horizontal gate width. All gate polygons must have an equal vertical length if they are not cut by GCUT. The vertical edge of the GCUT layer cannot lie inside or coincide with the gate layer, and it also cannot interact with the active region. The GCUT layer cannot exist without the gate layer.

The n- or p-select must always enclose the active region. The horizontal distance between two active areas varies if the diffusions have different or equal voltages. The SDT layer must always be inside the LISD layer, and it cannot be entirely outside of the active region. V0 should exactly be the same width as the M1 layer,

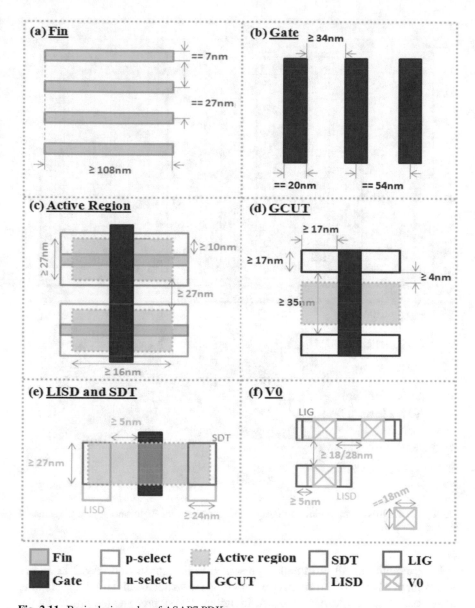

Fig. 2.11 Basic design rules of ASAP7 PDK

and it needs to be uniformly aligned for all vertical and horizontal directions. V0 must always interact with M1 layers and LISD/LIG. Moreover, this PDK requires a TAP cell in all layout designs to ensure the proper functionality of circuits. The TAP cell is responsible for connecting the back-gate of FinFET transistors. For FinFETs designed in an SG model, the back-gate has the same signal as the front-gate.

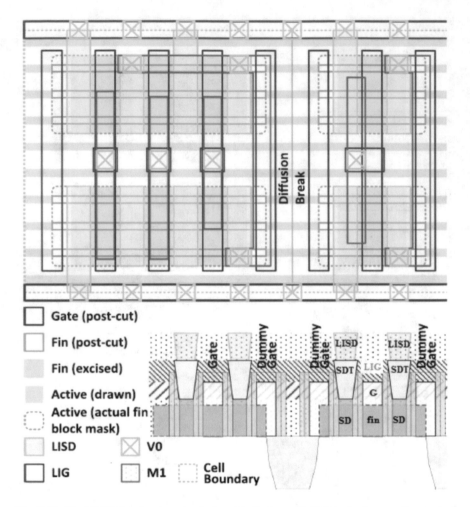

Fig. 2.12 The NAND3 and an inverter designed in the 7-nm FinFET technology on the standard cell template [15]

For exemplifying all information presented in this subsection, Fig. 2.12 shows the layout of a NAND3 and an inverter based on the standard cell template using the design rules from the ASAP7 PDK. Although the layout presents ten fins along the vertical axis, the transistors have only three fins. The excised fin rails represented in light gray, two between the active regions and two near to power, are necessary due to constraints imposed by FinFET technologies. However, they are not taken into account for the transistor sizing. The diffusions are connected using M1 and V0 contacts. As there is a diffusion break, each cell's active regions require a gate at either side (dummy gates). On the bottom of Fig. 2.12, the cross section view of cells is presented with detailed information about FEOL, MOL, and BEOL steps.

References

1. Agostinelli, M., Alioto, M., Esseni, D., Selmi, L.: Design and evaluation of mixed 3t-4t FinFET stacks for leakage reduction. In: Svensson, L., Monteiro, J. (eds.) Integrated Circuit and System Design. Power and Timing Modeling, Optimization and Simulation, pp. 31–41. Springer, Berlin (2009)
2. Alioto, M.: Analysis of layout density in FinFET standard cells and impact of fin technology. In: Proceedings of 2010 IEEE International Symposium on Circuits and Systems, pp. 3204–3207 (2010)
3. Alioto, M.: Comparative evaluation of layout density in 3t, 4t, and MT FinFET standard cells. IEEE Trans. Very Large Scale Integr. Syst. **19**(5), 751–762 (2011)
4. Anil, K.G., Henson, K., Biesemans, S., Collaert, N.: Layout density analysis of FinFETs. In: ESSDERC '03. 33rd Conference on European Solid-State Device Research, 2003, pp. 139–142 (2003)
5. Auth, C.: 22-nm fully-depleted tri-gate CMOS transistors. In: Proceedings of the IEEE 2012 Custom Integrated Circuits Conference, pp. 1–6 (2012)
6. Autran, J., Munteanu, D.: Soft Errors: From Particles to Circuits. Devices, Circuits, and systems. CRC Press, Boca Raton (2015)
7. Bhanushali, K., Davis, W.R.: Freepdk15: An open-source predictive process design kit for 15nm FinFET technology. In: Proceedings of the 2015 Symposium on International Symposium on Physical Design, p. 165–170. Association for Computing Machinery, New York (2015)
8. Bhattacharya, D., Jha, N.K.: FinFETs: from devices to architectures. Adv. Electron. **2014**, 1–21 (2014)
9. Borremans, J., Parvais, B., Dehan, M., Thijs, S., Wambacq, P., Mercha, A., Kuijk, M., Carchon, G., Decoutere, S.: Perspective of RF design in future planar and FinFET CMOS. In: 2008 IEEE Radio Frequency Integrated Circuits Symposium, pp. 75–78 (2008)
10. Boukortt, N., Hadri, B., Patanè, S., Caddemi, A., Crupi, G.: Electrical Characteristics of 8-nm SOI n-FinFETs. Springer, Berlin (2016)
11. Cartwright, J.: Intel Enters the Third Dimension (2011). https://www.nature.com/articles/news.2011.274
12. Chang, J.B., Guillorn, M., Solomon, P.M., Lin, C.., Engelmann, S.U., Pyzyna, A., Ott, J.A., Haensch, W.E.: Scaling of SOI FinFETs down to fin width of 4 nm for the 10nm technology node. In: 2011 Symposium on VLSI Technology - Digest of Technical Papers, pp. 12–13 (2011)
13. Chaudhuri, S., Mishra, P., Jha, N.K.: Accurate leakage estimation for FinFET standard cells using the response surface methodology. In: 2012 25th International Conference on VLSI Design, pp. 238–244 (2012)
14. Chauhan, Y.S., Lu, D.D., Vanugopalan, S., Khandelwal, S., Duarte, J.P., Paydavosi, N., Niknejad, A., Hu, C.: Chapter 11 - BSIM-CMG model parameter extraction. In: FinFET Modeling for IC Simulation and Design, pp. 231–243. Academic, Cambridge (2015)
15. Clark, L.T., Vashishtha, V., Shifren, L., Gujja, A., Sinha, S., Cline, B., Ramamurthy, C., Yeric, G.: ASAP7: A 7-nm FinFET predictive process design kit. Microelectron. J. **53**, 105–115 (2016)
16. Colinge, J.P. (ed.): FinFETs and Other Multi-Gate Transistors. Springer, Berlin (2008)
17. Conley, A.: FinFET vs. FD-SOI: Key Advantages & Disadvantages (2014). http://www.chipex.co.il/_Uploads/dbsAttachedFiles/ChipExAMAT.pdf
18. Cui, T., Li, J., Wang, Y., Nazarian, S., Pedram, M.: An exploration of applying gate-length-biasing techniques to deeply-scaled FinFETs operating in multiple voltage regimes. IEEE Trans. Emerg. Topics Comput. **6**(2), 172–183 (2018). https://doi.org/10.1109/TETC.2016.2640185

19. Cui, T., Xie, Q., Wang, Y., Nazarian, S., Pedram, M.: 7nm FinFET standard cell layout characterization and power density prediction in near- and super-threshold voltage regimes. In: International Green Computing Conference, pp. 1–7 (2014)
20. Dančak, C.: The FinFET: A Tutorial, pp. 37–69. Springer International Publishing, Cham (2018)
21. Datta, A., Goel, A., Cakici, R.T., Mahmoodi, H., Lekshmanan, D., Roy, K.: Modeling and circuit synthesis for independently controlled double gate FinFET devices. IEEE Trans. Comput. Aided Design Integr. Circuits Syst. **26**(11), 1957–1966 (2007)
22. Ding, Y., Chu, C., Mak, W.: Detailed routing for spacer-is-metal type self-aligned double/quadruple patterning lithography. In: 2015 52nd ACM/EDAC/IEEE Design Automation Conference (DAC), pp. 1–6 (2015)
23. Doris, B., Cheng, K., Khakifirooz, A., Liu, Q., Vinet, M.: Device design considerations for next generation cmos technology: Planar FDSOI and FinFET (invited). In: 2013 International Symposium on VLSI Technology, Systems and Application (VLSI-TSA), pp. 1–2 (2013)
24. Frank, D.J., Dennard, R.H., Nowak, E., Solomon, P.M., Taur, Y., Hon-Sum Philip Wong: Device scaling limits of Si MOSFETs and their application dependencies. Proc. IEEE **89**(3), 259–288 (2001)
25. Frumusanu, A.: TSMC Details 3nm Process Technology: Full Node Scaling for 2h22 Volume Production (2020). https://www.anandtech.com/show/16024/tsmc-details-3nm-process-technology-details-full-node-scaling-for-2h22
26. Ghaida, R.S., Agarwal, K.B., Nassif, S.R., Yuan, X., Liebmann, L.W., Gupta, P.: Layout decomposition and legalization for double-patterning technology. IEEE Trans. Comput. Aided Design Integr. Circuits Syst. **32**(2), 202–215 (2013)
27. Gu, J., Keane, J., Sapatnekar, S., Kim, C.H.: Statistical leakage estimation of double gate FinFET devices considering the width quantization property. IEEE Trans. Very Large Scale Integr. Syst. **16**(2), 206–209 (2008)
28. Guillorn, M., Chang, J., Bryant, A., Fuller, N., Dokumaci, O., Wang, X., Newbury, J., Babich, K., Ott, J., Haran, B., Yu, R., Lavoie, C., Klaus, D., Zhang, Y., Sikorski, E., Graham, W., To, B., Lofaro, M., Tornello, J., Koli, D., Yang, B., Pyzyna, A., Neumeyer, D., Khater, M., Yagishita, A., Kawasaki, H., Haensch, W.: FinFET performance advantage at 22nm: An AC perspective. In: 2008 Symposium on VLSI Technology, pp. 12–13 (2008)
29. Hibben, M.: TSMC, Not Intel, has the Lead in Semiconductor Processes (2018). https://seekingalpha.com/article/4151376-tsmc-not-intel-lead-in-semiconductor-processes
30. Hisamoto, D., Kaga, T., Kawamoto, Y., Takeda, E.: A fully depleted lean-channel transistor (delta)-a novel vertical ultra thin SOI MOSFET. In: International Technical Digest on Electron Devices Meeting, pp. 833–836 (1989)
31. Hisamoto, D., Lee, W.-C., Kedzierski, J., Takeuchi, H., Asano, K., Kuo, C., Anderson, E., King, T.-J., Bokor, J., Hu, C.: FinFET-a self-aligned double-gate mosfet scalable to 20 nm. IEEE Trans. Electron Devices **47**(12), 2320–2325 (2000)
32. Hu, C.: New sub-20nm transistors – why and how. In: 2011 48th ACM/EDAC/IEEE Design Automation Conference (DAC), pp. 460–463 (2011)
33. Intel: Intel's 10nm Technology: Delivering the Highest Logic Transistor Density in the Industry Through the Use of Hyper Scaling (2017). https://newsroom.intel.com/newsroom/wp-content/uploads/sites/11/2017/09/10-nm-icf-fact-sheet.pdf
34. ITRS: The International Technology Roadmap for Semiconductors (2011). http://www.itrs2.net/2011-itrs.html
35. James, D.: Intel to present on 22-nm Tri-gate technology at VLSI symposium (2012). https://sst.semiconductor-digest.com/chipworks_real_chips_blog/2012/04/12/intel-to-present-on-22-nm-tri-gate-technology-at-vlsi-symposium/
36. Kawa, J.: FinFET design, manufacturability, and reliability (2013). https://www.synopsys.com/designware-ip/technical-bulletin/FinFET-design.html
37. King, T.-J.: FinFETs for nanoscale CMOS digital integrated circuits. In: ICCAD-2005. IEEE/ACM International Conference on Computer-Aided Design, 2005., pp. 207–210 (2005)

38. Kleeberger, V.B., Graeb, H., Schlichtmann, U.: Predicting future product performance: Modeling and evaluation of standard cells in FinFET technologies. In: 2013 50th ACM/EDAC/IEEE Design Automation Conference (DAC), pp. 1–6 (2013)
39. Kuhn, K.J., Giles, M.D., Becher, D., Kolar, P., Kornfeld, A., Kotlyar, R., Ma, S.T., Maheshwari, A., Mudanai, S.: Process technology variation. IEEE Trans. Electron Devices 58(8), 2197–2208 (2011)
40. Kumar, V., Kirubaraj, A.: Submicron 70nm CMOS logic design with FinFETs. Int. J. Eng. Sci. Technol. 2, 4751–4758 (2010)
41. Liu, Y., Li, Y.: Aspherical surfaces design for extreme ultraviolet lithographic objective with correction of thermal aberration. Optical Eng. 55(9), 1–6 (2016)
42. McLellan, P.: FinFET Custom Design (2014). https://semiwiki.com/semiconductor-manufacturers/tsmc/3327-FinFET-custom-design/
43. Meinhardt, C.: Variabilidade em FinFETs. Thesis (Doutorado em Ciência da Computação) – Instituto de Informática - UFRGS (2014)
44. Mentor: As Nodes Advance, So Must Power Analysis (2014). https://semiengineering.com/as-nodes-advance-so-must-power-analysis/
45. Mishra, P., Muttreja, A., Jha, N.K.: Nanoelectronic Circuit Design, pp. 23–54. Springer, New York (2011)
46. Moore, G.E.: Cramming more components onto integrated circuits, reprinted from electronics, volume 38, number 8, april 19, 1965, pp.114 ff. IEEE Solid-State Circuits Soc. Newslett. 11(3), 33–35 (2006)
47. Muttreja, A., Agarwal, N., Jha, N.K.: CMOS logic design with independent-gate FinFETs. In: 2007 25th International Conference on Computer Design, pp. 560–567 (2007)
48. Nowak, E.J., Aller, I., Ludwig, T., Kim, K., Joshi, R.V., Ching-Te Chuang, Bernstein, K., Puri, R.: Turning silicon on its edge [double gate CMOS/FinFET technology]. IEEE Circuits Devices Mag. 20(1), 20–31 (2004)
49. Nowak, E.J., Rainey, B.A., Fried, D.M., Kedzierski, J., Ieong, M., Leipold, W., Wright, J., Breitwisch, M.: A functional FinFET-DGCMOS sram cell. In: Digest. International Electron Devices Meeting, pp. 411–414 (2002)
50. Park, J.-T., Colinge, J., Diaz, C.H.: Pi-gate SOI MOSFET. IEEE Electron Device Lett. 22(8), 405–406 (2001)
51. Poljak, M., Jovanovic, V., Suligoj, T.: SOI vs. bulk FinFET: Body doping and corner effects influence on device characteristics. In: MELECON 2008 - The 14th IEEE Mediterranean Electrotechnical Conference, pp. 425–430 (2008)
52. Posser, G., Belomo, J., Meinhardt, C., Reis, R.: Performance improvement with dedicated transistor sizing for mosfet and FinFET devices. In: 2014 IEEE Computer Society Annual Symposium on VLSI, pp. 418–423 (2014)
53. Pradhan, K.P., Sahu, P.K., Ranjan, R.: Investigation on asymmetric dual-k spacer (ads) trigate wavy FinFET: A novel device. In: 2016 3rd International Conference on Devices, Circuits and Systems (ICDCS), pp. 137–140 (2016)
54. PTM: Predictive Technological Model (2012). http://ptm.asu.edu/
55. Rainey, B.A., Fried, D.M., Ieong, M., Kedzierski, J., Nowak, E.J.: Demonstration of FinFET CMOS circuits. In: 60th DRC. Conference Digest Device Research Conference, pp. 47–48 (2002)
56. Ranjan, A.: Physical Verification of FinFET and FD-SOI Devices (2013). https://www.techdesignforums.com/practice/technique/physical-verification-design-FinFET-fd-soi/
57. Ranjan, A.: Micro-Architectural Exploration for Low Power Design (2015). https://semiengineering.com/micro-architectural-exploration-for-low-power-design/
58. Rieger, M.L.: Communication theory in optical lithography. J. Micro/Nanolith. MEMS MOEMS 11(1), 1–11 (2012)
59. Rostami, M., Mohanram, K.: Dual-v_{th} independent-gate FinFETs for low power logic circuits. IEEE Trans. Computer-Aided Design Integr. Circuits Syst. 30(3), 337–349 (2011)

60. Roy, K., Mahmoodi, H., Mukhopadhyay, S., Ananthan, H., Bansal, A., Cakici, T.: Double-gate SOI devices for low-power and high-performance applications. In: 19th International Conference on VLSI Design Held Jointly with 5th International Conference on Embedded Systems Design (VLSID'06), pp. 8 pp.– (2006)
61. Saha, R., Bhowmick, B., Baishya, S.: Si and Ge step-FinFETs: work function variability, optimization and electrical parameters. Superlattices Microstruct. **107**, 5–16 (2017)
62. Sairam, T., Zhao, W., Cao, Y.: Optimizing FinFET technology for high-speed and low-power design. In: Proceedings of the 17th ACM Great Lakes Symposium on VLSI, GLSVLSI '07, p. 73–77. Association for Computing Machinery (2007)
63. Samsung: Samsung Launches Premium Exynos 9 Series Processor Built on the World's First 10nm FinFET Process Technology (2017). https://news.samsung.com/global/samsung-launches-premium-exynos-9-series-processor-built-on-the-worlds-first-10nm-FinFET-process-technology
64. Seifert, N., Jahinuzzaman, S., Velamala, J., Ascazubi, R., Patel, N., Gill, B., Basile, J., Hicks, J.: Soft error rate improvements in 14-nm technology featuring second-generation 3d tri-gate transistors. IEEE Trans. Nuclear Sci. **62**(6), 2570–2577 (2015)
65. Sekigawa, T., Hayashi, Y.: Calculated threshold-voltage characteristics of an XMOS transistor having an additional bottom gate. Solid State Electron. **27**(8–9), 827–828 (1984)
66. Sicard, E.: Introducing 14-nm FinFET Technology in Microwind (2017). https://hal.archives-ouvertes.fr/hal-01541171/document
67. Simsir, M.O., Bhoj, A., Jha, N.K.: Fault modeling for FinFET circuits. In: 2010 IEEE/ACM International Symposium on Nanoscale Architectures, pp. 41–46 (2010)
68. Singh, N., Agarwal, A., Bera, L.K., Liow, T.Y., Yang, R., Rustagi, S.C., Tung, C.H., Kumar, R., Lo, G.Q., Balasubramanian, N., Kwong, D..: High-performance fully depleted silicon nanowire (diameter /spl les/ 5 nm) gate-all-around CMOS devices. IEEE Electron Device Lett. **27**(5), 383–386 (2006)
69. Sinha, S., Yeric, G., Chandra, V., Cline, B., Cao, Y.: Exploring sub-20nm FinFET design with predictive technology models. In: DAC Design Automation Conference 2012, pp. 283–288 (2012)
70. Skotnicki, T., Hutchby, J.A., Tsu-Jae King, Wong, H..P., Boeuf, F.: The end of CMOS scaling: toward the introduction of new materials and structural changes to improve mosfet performance. IEEE Circuits Devices Mag. **21**(1), 16–26 (2005)
71. Solomon, P.M., Guarini, K.W., Zhang, Y., Chan, K., Jones, E.C., Cohen, G.M., Krasnoperova, A., Ronay, M., Dokumaci, O., Hovel, H.J., Bucchignano, J.J., Cabral, C., Lavoie, C., Ku, V., Boyd, D.C., Petrarca, K., Yoon, J.H., Babich, I.V., Treichler, J., Kozlowski, P.M., Newbury, J.S., D'Emic, C.P., Sicina, R.M., Benedict, J., Wong, H..P.: Two gates are better than one [double-gate mosfet process]. IEEE Circuits Devices Mag. **19**(1), 48–62 (2003)
72. Subramaniana, V., Parvais, B., Borremans, J., Mercha, A., Linten, D., Wambacq, P., Loo, J., Dehan, M., Collaert, N., Kubicek, S., Lander, R.J.P., Hooker, J.C., Cubaynes, F.N., Donnay, S., Jurczak, M., Groeseneken, G., Sansen, W., Decoutere, S.: Device and circuit-level analog performance trade-offs: A comparative study of planar bulk FETs versus FinFETs. In: IEEE International Electron Devices Meeting, 2005. IEDM Technical Digest., pp. 898–901 (2005)
73. Swahn, B., Hassoun, S.: Gate sizing: FinFETs vs 32nm bulk MOSFETs. In: 2006 43rd ACM/IEEE Design Automation Conference, pp. 528–531 (2006)
74. Tang, S.H., Chang, L., Lindert, N., Choi, Y.-K., Lee, W.-C., Huang, X., Subramanian, V., Bokor, J., King, T.-J., Hu, C.: FinFET-a quasi-planar double-gate MOSFET. In: 2001 IEEE International Solid-State Circuits Conference. Digest of Technical Papers. ISSCC (Cat. No.01CH37177), pp. 118–119 (2001)
75. Trivedi, V.P., Fossum, J.G., Zhang, W.: Threshold voltage and bulk inversion effects in nonclassical CMOS devices with undoped ultra-thin bodies. Solid State Electron. **51**(1), 170–178 (2007)

76. Wambacq, P., Verbruggen, B., Scheir, K., Borremans, J., De Heyn, V., Van der Plas, G., Mercha, A., Parvais, B., Subramanian, V., Jurczak, M., Decoutere, S., Donnay, S.: Analog and RF circuits in 45 nm CMOS and below: Planar bulk versus FinFET. In: 2006 Proceedings of the 32nd European Solid-State Circuits Conference, pp. 54–57 (2006)

77. Wilson, D., Hayhurst, R., Oblea, A., Parke, S., Hackler, D.: Flexfet: Independently-double-gated SOI transistor with variable Vt and 0.5 V operation achieving near ideal subthreshold slope. In: 2007 IEEE International SOI Conference, pp. 147–148 (2007)

78. Wimer, S.: Planar CMOS to multi-gate layout conversion for maximal fin utilization. Integration **47**(1), 115–122 (2014)

79. Yang, F.L., Chen, H.Y., Chen, F.C., Huang, C.C., Chang, C.Y., Chiu, H.K., Lee, C.C., Chen, C.C., Huang, H.T., Chen, C.J., Tao, H.J., Yeo, Y.-C., Liang, M.-S., Hu, C.: 25nm CMOS omega FETs. In: Digest. International Electron Devices Meeting, pp. 255–258 (2002)

80. Yu, B., Chang, L., Ahmed, S., Wang, H., Bell, S., Yang, C.-Y., Tabery, C., Ho, C., Xiang, Q., King, T.-J., Bokor, J., Hu, C., Lin, M.-R., Kyser, D.: FinFET scaling to 10 nm gate length. In: Digest. International Electron Devices Meeting, pp. 251–254 (2002)

81. Zarei, M.Y., Asadpour, R., Mohammadi, S., Afzali-Kusha, A., Seyyedi, R.: Modeling symmetrical independent gate FinFET using predictive technology model. In: Proceedings of the 23rd ACM International Conference on Great Lakes Symposium on VLSI, GLSVLSI '13, p. 299–304. Association for Computing Machinery (2013)

82. Zhang, B.: FinFET standard cell optimization for performance and manufacturability. UT Electronic Theses and Dissertations (2012)

83. Zhang, W., Fossum, J.G., Mathew, L., Du, Y.: Physical insights regarding design and performance of independent-gate FinFETs. IEEE Trans. Electron Devices **52**(10), 2198–2206 (2005)

84. Zhao, W., Cao, Y.: New generation of predictive technology model for sub-45nm design exploration. In: 7th International Symposium on Quality Electronic Design (ISQED'06), pp. 6 pp.–590 (2006)

Chapter 3
Reliability Challenges in FinFETs

The technology scaling and the adoption of FinFET devices brought several benefits, but some challenges were also introduced. The quantization feature to increase the transistor width imposed restrictions at the layout level because the circuits need to be always designed into a grid, reducing the design flexibility. First, this chapter gives a general overview of the reliability challenges in FinFET nodes. After, the focus is mainly on details of the main sources and classifications of process variations and radiation-induced soft errors. Additionally, some related works that evaluate and mitigate the effects of reliability challenges are presented.

3.1 Overview

FinFET devices require careful resistances and capacitances modeling with satisfactory tools to realize the RC extraction, avoiding inappropriate device characterization and circuit performance degradation. The Miller effect influences the reliability of the circuit and reinforces the need for power and timing analysis accurately [69].

Examples of the most common FinFET problems are the fringe capacitance to contact/facet, low-k spacer, fin/gate fidelity, contact resistances, chemical–mechanical planarization (CMP) polish, threshold voltage tuning, susceptibility to process variability, fin strain engineering, quantization feature to transistor sizing, and the surface orientation [49]. Moreover, the circuits become more susceptible to transient faults from space and terrestrial radiations and permanent events. Figure 3.1 summarizes the FinFET challenges, but all of them raise essential topics related to the reliability of electronic systems that need to be better investigated.

According to [16], the reliability in FinFET technologies can be divided into static and dynamic sources, as Fig. 3.2 illustrates. Static sources are usually random, permanent in time, and immediately noticeable after the manufacturing process.

© The Author(s), under exclusive license to Springer Nature Switzerland AG 2021
A. Zimpeck et al., *Mitigating Process Variability and Soft Errors at Circuit-Level for FinFETs*, https://doi.org/10.1007/978-3-030-68368-9_3

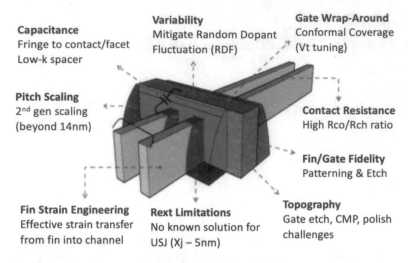

Fig. 3.1   Some challenges for FinFET technologies [49]

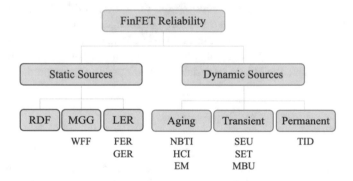

Fig. 3.2   Examples of static and dynamic sources in FinFET devices

Examples of static variation sources are random dopant fluctuation (RDF), line edge roughness (LER), and metal gate granularity (MGG).

On the other hand, dynamic sources are time-varying, suffering modifications according to operating conditions like temperature oscillation, supply voltage drop, switching activity, environmental noise, and radiation exposure. Examples of dynamic sources are the aging effects (e.g., Bias-Temperature Instability [BTI], Hot-Carrier Injection [HCI], and Electromigration [EM]), transient faults (e.g., Single Event Upset [SEU], Single Event Transient [SET], and Multiple-Bit Upset [MBU]), and permanent events (e.g., Total Ionizing Dose [TID]).

Currently, process variability and radiation-induced soft errors are considered two of the significant reliability challenges for commercial electronic systems manufactured in FinFET technologies. From a design standpoint, considerable efforts should be made to reduce the impacts introduced by these issues [86, 97]. In this way, the next subsections focus on evaluating the sources and types of process variations and radiation effects, explaining the impact on integrated circuits.

3.2 Process Variability

Variability is related to the random deviation, which causes an increase or decrease of typical design specifications [97]. The main issue associated with variability is the uncertainty about the correct circuit operation because there is no guarantee that a circuit will behave as expected after the manufacturing process. Each circuit can present a different electrical behavior, such as abnormal power consumption and performance deviation due to the variability. The unexpected behavior due to variations can stimulate the circuit degradation besides making it inappropriate for their initial purpose [88]. The variability sources can be divided into three main categories: environmental, reliability, and physical [82]. The first two categories are dynamic (time-varying), while the last one is static, occurring during the manufacturing process.

Environmental factors are deviations in the operating conditions during the circuit's lifetime due to architectural and operational decisions such as the design of power lines and cell placement. The most common examples of this category are the oscillations in switching frequency, temperature, supply voltage, and environmental noise. The supply voltage is variable in a chip, and voltage drops occur mainly due to non-zero resistances in the power supply networks [70]. Supply voltage (V_{DD}) has a quadratic relationship with dynamic power, according to Eq. 3.1, where f is the switching frequency, and C is the capacitance between the output nodes. This relation becomes an attractive solution for low-power applications, but it has to be used carefully because the clock frequency is reduced significantly.

$$P_{dynamic} = f \times C \times VDD^2 \tag{3.1}$$

Moreover, the supply voltage deviation also affects the propagation delays due to the transistor saturation current. This association is exponential for a wide voltage range. Temperature oscillations can compromise interconnections and the behavior of electronic components in a chip. Higher heat flux results in higher temperature, creating hot spots, which generate temperature variations across the die. At high temperatures, devices may not meet performance requirements due to threshold voltage drift, higher leakage currents, and propagation delay. Moreover, temperature variations across communicating blocks on the same chip may cause logic or functional failures [16]. However, the structure of multigate devices has diminished the thermal conductivity due to the small and confined dimensions of the fin [63, 125].

Reliability factors are related to the transistor aging due to the rising electrical fields through the oxide thickness presented in modern circuits. NBTI, HCI, and electromigration are classical problems in this category. NBTI is a degradation

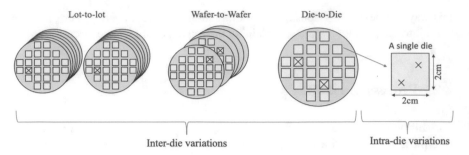

Fig. 3.3 Intra-die and inter-die variations [23]

factor that negatively affects the performance and noise margins because the electrical fields generate interface traps in p-type devices, which results in the unwanted increase in the threshold voltage [57]. HCI is another type of degradation mechanism, and it arises from the heating inside the transistor channel during the circuit operation. Electromigration causes shorts and opens in metal interconnects, leading to interconnection failures decreasing the mean time to failure (MTTF) of the chip at sub-45nm nodes [91].

Physical factors are associated with structural or doping variations after the manufacturing process. They can be classified into intra-die and inter-die variations [79], as shown in Fig. 3.3. The intra-die variations are deviations inside the same chip, in different locations of the same die. The inter-die variations are characterized by die-to-die, wafer-to-wafer, and lot-to-lot fluctuations. The die-to-die variations occur between two dies in the same wafer. The wafer-to-wafer variations happen between dies in different wafers, and the lot-to-lot variations appear between dies in different lots of production volume.

Integrated circuits with manufacturing defects or deviations from initial spec-ifications should be discarded. The yield loss in the semiconductor industry can be divided into catastrophic and parametric aspects [46]. The former is related to functional failures such as open trails or short circuits, making the circuit not work correctly. Figure 3.4a shows an example where a bridging fault on metal 3 occurs, changing the circuit's behavior. On the other hand, the parametric yield loss is caused by process variations where a chip is functionally correct, but it fails to meet some power or performance criteria. Figure 3.4b shows a poly layer with some geometric deviations. The manufacturing process's high costs are directly related to the variability effects due to the many redesign steps required until the expected behavior is achieved. For example, a chip designed in planar CMOS technology was modified around four times before reaching the production volume [104]. Still, this number is even more critical for designs that employ sub-22nm technologies.

The manufacturing process introduces variability at each stage. Once ready, a chip is also susceptible to several natural changes caused by the environment over

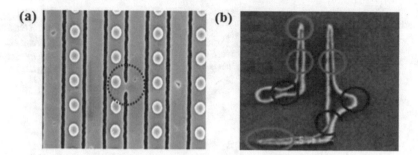

Fig. 3.4 (**a**) Catastrophic and (**b**) parametric yield losses [46, 62]

its lifetime. Environmental factors increase the dynamic variability, and the process variations intensify the static variability. In this way, dealing with variability is an increasingly complex problem. The challenges imposed by variability require new design methodologies and new electronic design automation (EDA) tools to predict and minimize their effects on the integrated circuits. Moreover, researchers highlight that it is no longer sufficient to focus only on the deviations of the threshold voltage in the design development considering FinFET technologies. It is also necessary to evaluate all variations in the electrical characteristics such as I_{ON}/I_{OFF} currents, power consumption, and performance.

3.2.1 Sources of Process Variability

The technology scaling to maintain the pace of performance and density gains results in an increase of design complexity with more potential variability sources. In this way, sub-22nm technology nodes tend to be more susceptible to process variability effects. The process variability arises from the inaccuracy wavelength used to transfer the small geometric patterns to the wafer, the use of high-k dielectrics to improve the gate control on the channel region, or the alteration in the doping density.

The most significant process variability sources are the line edge roughness, metal gate granularity, and random dopant fluctuation. The LER can be subdivided into fin edge roughness (FER) and gate edge roughness (GER). Figure 3.5 shows an overview of how and where these sources modify the transistor structure during the manufacturing process. These variations can compromise the entire blocks of cells besides reducing the performance and energy efficiency of the chip. The individual contributions of each source are process dependent. The combined effect can impact the device behavior and the parametric yield loss [2]. More detailed information about variability sources in FinFET devices will be explored in the next subsections.

Fig. 3.5 Major random
variation sources in FinFETs
[55]

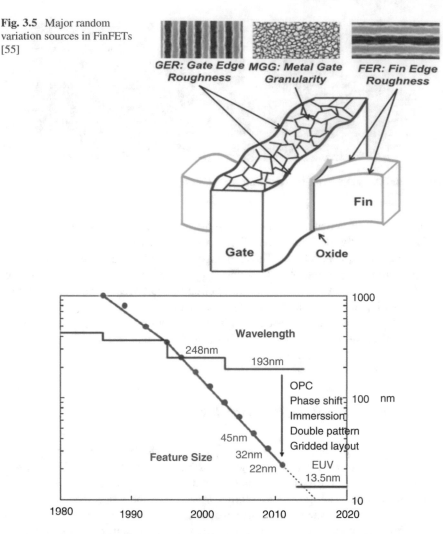

Fig. 3.6 Relationship between technology scaling and wavelength into the lithography step [95]

Line Edge Roughness

Typically, the lithography step in the fabrication process uses ultraviolet light
to transfer geometric shapes to thin silicon slices. However, the wavelength is
a property of the light source that has not kept pace with the device scaling.
The transition of light wavelength from 193 nm to 13.5 nm is a slow procedure,
as shown in Fig. 3.6, becoming harder the use of 193-nm lithography as chip
manufacturing advances more in-depth into the nanometer regime due to the circuit
design complexity, increased clock frequency, and edge placement errors [12].

Fig. 3.7 Geometric
parameters for FinFET
devices [40]

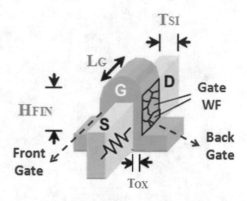

In this way, circuits designed in FinFET technologies are more prone to suffer
from the LER phenomenon due to the small standards imposed by sub-22nm nodes.
Line edge roughness corresponds to a deviation in the transistor edge position
compared with the best fit line of the ideal shape. The imperfections caused by the
optical performance loss can influence the fin height (H_{FIN}), fin thickness (T_{SI}),
the gate length (L_G), and oxide thickness (T_{OX}), as shown in Fig. 3.7 [40].

The deviations in the gate region are classified as GER, while the variations in the
fin area are called FER. Gate length, fin thickness, and oxide thickness were pointed
as the central sources of threshold voltage variability for FinFET devices [40]. FER
deviations introduced higher variability on the on-state current since FER affects the
channel and also the source/drain resistances [118]. The impact on I_{ON} is a little
more meaningful when the fin height suffers deviations instead of the fin thickness.
The off-state current suffers more deviations when the gate length is modified during
the fabrication process [72]. The impact of oxide thickness variations in FinFET
devices can be considered negligible when on- and off-state currents were observed
[131].

To reduce the LER, the lithographers use many approaches to bypass the features
much smaller than were allowed by the resolution criteria of 193-nm lithography,
such as optical proximity correction. Moreover, some layers replaced the single
exposure by multiple patterning methodologies to provide enough resolution in
the integrated circuits. Figure 3.8 shows the sub-45nm technology advancement
regarding the use of single, double, and quadruple patterning for the main layers.
Usually, the fin layers are implemented using quadruple patterning such as the
SAQP method [28]. The complexity of multiple patterning is because the regular
mask is broken up into four incremental mask levels (or two in double patterning).
Each part has process variations that do not correlate with one another. It is
necessary to align the mask layers of additional exposures accurately on top of each
other for better patterning quality, obeying the overlay requirements. Furthermore,
multiple patterning methodologies impose new physical verification at the layout
level.

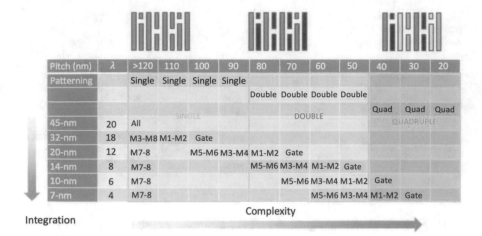

Pitch (nm)	λ	>120	110	100	90	80	70	60	50	40	30	20
Patterning		Single	Single	Single	Single							
						Double	Double	Double	Double			
										Quad	Quad	Quad
45-nm	20	All										
32-nm	18	M3-M8	M1-M2	Gate								
20-nm	12	M7-8		M5-M6	M3-M4	M1-M2	Gate					
14-nm	8	M7-8					M5-M6	M3-M4	M1-M2	Gate		
10-nm	6	M7-8						M5-M6	M3-M4	M1-M2	Gate	
7-nm	4	M7-8							M5-M6	M3-M4	M1-M2	Gate

Integration   Complexity

Fig. 3.8 Single, double, and quadruple patterning applied in the layers of advanced technology nodes [17]

Metal Gate Granularity

MGG gained prominence since the adoption of high-k dielectrics to improve the gate control on the channel region in sub-45nm technologies. The energy difference between the vacuum level and the Fermi level of a solid is called work-function. The metal WF is the minimum energy required to move an electron from the Fermi level to the vacuum level, and it depends on the type of metal used. MGG refers to the random orientation of the different metal grains leading to variations in the gate work-function. According to Fig. 3.9, in the ideal fabrication process, metal gate devices have the gates produced with a unique metal uniformly aligned.

Nevertheless, in the real fabrication process, metal gate devices are generally produced using metals with different work-functions (ϕm) randomly aligned, which implies higher work-function fluctuation (WFF) [33]. Wang et al. highlight the high correlation between the variability in the on-state current and threshold voltage under metal gate granularity [118]. In general, multigate devices exhibit higher MGG variability compared to other process variability phenomena like LER and RDF. Previous research indicates that work-function fluctuation is the main source of variability in FinFET devices [72].

Random Dopant Fluctuation

RDF arises from the variation in the implanted impurity concentration modifying the discreteness of dopant atoms in the transistor channel [15]. The change in the number or placement of dopant atoms results in threshold voltage deviations that

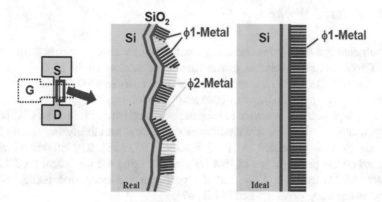

Fig. 3.9 Metal gate alignment in a real and ideal manufacturing process [88]

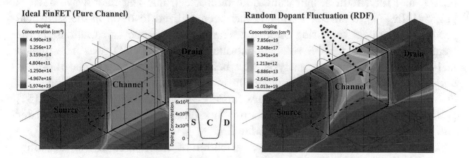

Fig. 3.10 Ideal FinFET versus a device with random dopant fluctuations [105]

directly affect the transistor properties. The variation in the source/drain resistance due to doping alterations dominates the RDF variability incurring in the absence of channel doping [118]. Since the RDF is a local form of physical variability, neighboring transistors can present different dopant concentrations.

The technology scaling has been adjusting the channel doping to meet the targeted threshold voltage and on- and off-state current expectations. Moreover, to control the leakage currents and other challenges in bulk CMOS technologies, the total number of channel dopants increased significantly, resulting in a larger variety of dopant atoms. For this reason, the RDF was considered the major source of variability in planar technologies [83]. As the fin thickness is reduced, the FinFET becomes fully depleted. The channel can be lightly doped, which provides a significant reduction in the threshold voltage fluctuations due to RDF. It implies that FinFET devices suffer less from dopant-induced variations [1, 55], improving the manufacturing yield. Figure 3.10 shows on the left a FinFET with the nominal doping concentration, i.e., without RDF influence. On the right, the local doping concentration is non-uniform along with the source–channel–drain region of the device.

3.2.2 Related Works

This subsection summarizes the main related works about process variability in FinFET technologies available in the literature. A case study of Gold Standard Simulation (GSS) simulated some sources of variability in a 3D FinFET model. The experiments evaluated the impact of process variations on the V_{TH}, I_{ON}, and I_{OFF} analyzing the electrical behavior in subthreshold and saturation regions [45]. Basic concepts, characterization, and challenges of process variability in digital circuits and systems were presented in [15, 33, 40, 82, 97, 118]. The effects of random variations on the performance of FinFET circuits using the analytical models were evaluated in [1, 20, 112, 119], and the performance estimation using response surface methodology was studied in [31, 47].

Artificial neural networks have been widely used as a computer-aided design (CAD) tool for circuit design using both microelectronic and microwave devices [11, 54]. In [79], a similar technique has been utilized to develop the statistical performance model for a ring oscillator. The impact of process variations on the frequency of a bulk CMOS voltage-controlled oscillator (VCO) was investigated in [44] using Monte Carlo analysis. The behavior of SRAM cells and flip-flops under process variability was explored in [42, 84, 92]. Mitigation techniques for environmental and reliability variations in FinFET devices were proposed in [53] [120]. A method to model the effects of work-function variations in a transistor with a high-k metal gate was presented in [2].

The quantitative evaluation of the contribution of different statistical variability sources is provided in [22] for a low-power bulk NFET. In [111], FinFET gate sizing was formulated as a power minimization subject to delay, considering the operating temperature and weighting factors as constraints. One methodology to find the optimal sizing of FinFET circuits under process variations optimizing the worst-case cost for a given yield requirement was presented in [61]. In [110], the sizing problem is formulated as a mixed-integer non-linear program (MINLP) to minimize the area of FinFET devices.

The influence of process variability on I_{ON} and I_{OFF} currents of PFET and NFET transistors was analyzed for a set of predictive FinFET technologies from 20 nm to 7 nm [71]. Results showed that fin height has a small standard deviation, while gate length and fin thickness have a considerable difference from nominal conditions. However, for all setups evaluated, work-function fluctuation showed to be the most impacted parameter under the process variability effects with large standard deviation results.

Additional research was done in [131], focusing on analyzing the impact of temperature variations on 20-nm FinFET devices. LSTP devices are more sensitive to temperature oscillations. PFET devices are more impacted by the temperature, with an increase of 7.27 µA and 7.82µA in the I_{ON} to HP and LSTP devices, respectively. LSTP devices are up to 25% more susceptible to temperature variations. Another point is that PFET devices are approximately 30% more sensitive to temperature effects.

An evaluation of the process, voltage, and temperature (PVT) variations impact on total/static power and timing in a set of cells was made in [72] and [126]. Under voltage variations, some cells presented up to 70% of power-delay-product (PDP) reduction. However, voltage reduction provokes a timing increase by more than three times. Total power consumption is the principal parameter impacted by the temperature increase. At higher temperatures, the power increase can reach results of around five times the nominal values. Finally, WFF variation has a significant impact on I_{OFF} and the static power consumption of standard cells. For cells with a similar function but with a different number of inputs, it is possible to note a decreasing WFF sensibility as the number of inputs rises.

The impact of PVT variations and NBTI/PBTI aging on the write noise margins is measured and compared for a set of MOSFET and FinFET flip-flops [57]. The authors adopted the 16-nm FinFET predictive technology from PTM to obtain the results. The smaller standard deviation in FinFET cells results in better performance for write failure probability at a given input voltage noise. Supply voltage dependence of noise margins tends to be always linear and poorly affected by aging. On the other hand, the temperature dependence of noise margins is linear, with opposite behavior in MOSFET and FinFET cells.

Different sources of process variability and their impact on FinFET-based logic cells were explored in [56]. Both TCAD and PTM device models were used and compared concerning the performance metrics of the NAND2 and NOR2 gates adopting the 14-nm technology node. They conclude that LER and MGG are the dominating local variability sources affecting the gate delay, while the RDF has a minor role. Otherwise, since the threshold voltage is highly sensitive to RDF, it has a dominant impact on leakage power variation along with the LER. There is a threshold voltage difference between the device models, and then, the deviations are of an order of magnitude higher for PTM.

The impact of fin shape variability on the short channel effect control is investigated through TCAD simulations with both 14-nm and 10-nm FinFET nodes [113]. This work reveals that fin height and fin thickness variations besides the taper angle significantly affect the electrostatics of the device. Results showed a PFET transistor under this three-fin shape variation and verified the impact in the I_{ON}/I_{OFF} currents. Compared to the nominal case, the higher, narrower, and more tapered fins show less deviation on I_{OFF} currents. They suggested that the fin thickness and angle variability suppression would be essential to guarantee the variability robustness, especially in advanced nodes.

The impact of PVT variations on performance and power consumption considering different transistor sizing techniques applied to a fixed subset of gates was presented in [129] and [130]. The transistor sizing techniques analyzed were minimum transistor sizing, which corresponds to all cells with the number of fins equal to 1, the logical effort (LE), and the optimized transistor sizing using the sizing presented in [90]. Results point out that transistor sizing regarding the process variability is not a trivial choice. The most indicated technique is the OTS for FinFET cells, but it presents a large area overhead.

A novel FinFET structure with body spacers was proposed to improve the FER variation produced during the manufacturing process [121]. The effective H_{FIN} with body spacers is precisely controlled because it only depends on the silicon epitaxy layer thickness. Device simulation using the Sentaurus TCAD demonstrated an improvement from 33.46% to 8.05% in the I_{ON} current variation when the body spacers are applied in 10-nm bulk n-FinFET transistors. The gain is even more significant for devices when higher body spacer heights were used. Moreover, manufacturing FinFET with body spacers needs no extra lithography step becoming a promising mitigation technique for the industrial community.

A report about the impact of device scaling on the performance of a FinFET device due to gate work-function fluctuation and random dopant fluctuation was done in [83]. 3D device simulation considering the technology nodes of 14 nm, 10 nm, and 7 nm was performed. The WFF and RDF variations (observing standard deviation) of both threshold voltage and subthreshold swing are significant. Their investigation reveals that the impact of RDF can be reduced without to alter the channel doping, but meeting the targeted $V_{TH}/I_{OFF}/I_{ON}$. However, the negative side is the increase of the relative impact of WFF as the technology scaling down.

The threshold voltage variability induced by WFF for different grain sizes (10, 7, and 5 nm) in a 14-nm FinFET technology is analyzed using 3D simulations [94]. They observed that with a reduction in grain size, the threshold voltage variations decrease linearly. They have seen that the different fin shapes have around of 6% shift in the threshold voltage where there is approximately 16% improvement in the standard deviation of the V_{TH}. Furthermore, reducing the average grain size from 10 nm down to 5 nm results in an approximately 45% reduction in variability induced by WFF.

A detailed set of predictive data about the behavior of FinFET and Trigate devices considering process variability effects in I_{ON} and I_{OFF} currents was provided in [127] and [128]. Process variation analysis considers the individual contribution of the main geometric parameters of the devices using the predictive sub-22nm technologies from PTM-MG. Individually, L_G, H_{FIN}, and W_{FIN} parameters slightly affect the I_{ON} currents considering geometric deviations in the range of 5–20%. On the other hand, the I_{OFF} suffers the higher impact of geometric variability, mainly on FinFET devices. PFET devices and the LSTP model are also more sensitive than NFET devices and high-performance models. The results highlight that Trigate devices are up to 10% less sensitive to gate length variations.

A simple device-level characterization approach to quantitatively evaluate the impacts of different random variation sources in FinFETs is proposed in [55]. The variations of threshold voltage induced by LER and MGG are theoretically decomposed based on the distinction in physical mechanisms and their influences on different electrical characteristics. The effectiveness of the proposed method was confirmed through both TCAD simulations and experimental results. There is a considerable increase when L_G shrinks, while for MGG variations, V_{TH} remains consistent. This work can provide helpful guidelines for variation-aware technology development.

Two-step FinFET devices with different fin materials (Si and Ge) were analyzed under WFF and geometric variations compared to conventional FinFET [98]. The parametric analysis showed that Si step-FinFET is more immune to subthreshold swing, drain induced barrier lowering, and threshold voltage. At the same time, Ge step-FinFET has a higher I_{ON}/I_{OFF} ratio and a lower intrinsic delay at different length and oxide thickness. When the gate metal work-function fluctuation is inserted, Si step-FinFET presented a minor variation in the threshold voltage and subthreshold swing, but a higher variation in the I_{ON}/I_{OFF} ratio than in conventional FinFET. The proposed device performs better in low-power applications.

Temperature dependence is of utmost importance for the performance and power dissipation analysis. The temperature dependence of bulk double-gate FinFET and Trigate MOSFET devices is investigated in [3]. Additionally, it is also evaluated the study for the zero temperature coefficient (ZTC) condition. The results indicate that the increase in leakage current can reach more than 40 times compared to the nominal temperature for high-performance applications. Trigate devices have shown to be more sensitive to these variations with a difference of up to 19.7% in I_{OFF} current compared to FinFETs.

The impact of oxide thickness on threshold voltage variation induced by WFF in multigate devices was investigated using 3D simulation [64]. The WFF-induced threshold voltage variation does not significantly vary with dielectric material but increases with decreasing physical oxide thickness. The electric field tends to be locally concentrated, causing a considerable deviation of electrostatic potential as T_{OX} becomes thinner. They conclude that it is possible to alleviate the WFF-induced V_{TH} variation without significant performance degradation if the gate dielectric layer becomes thicker with appropriately adopted higher-k engineering.

The effects of fin thickness scaling of p- and n-type 10-nm FinFET and the correlation of the WFF with the electrical performance of the devices were investigated in [89]. They observed that the transfer characteristics are increased drain current in the linear region towards increased T_{SI} for both p- and n-FinFETs. The threshold voltage is shifted to the right for p-type as the work-function is increased. Oppositely, for n-type, they turned to the left as the work-function reduced. The I_{ON}/I_{OFF} ratio for the low-performance device shows that the magnitude drops to 63% and 82% in n- and p-type, respectively, when the fin width is changed from 4 nm to 8 nm.

A 3D simulation study to evaluate the threshold voltage variability induced by statistical parameter fluctuations in 14-nm bulk and SOI FinFET structures was done in [93]. They have studied and explored the influence of various statistical variability sources such as RDF, oxide thickness variation, and WFF on threshold voltage performance for both bulk and SOI FinFET structures. The simulation results suggest that the threshold voltage variability in the SOI FinFET structure shows 32% improvement compared to the bulk FinFET structure.

In [35], they studied the effects of two essential variation sources such as work-function fluctuation of the gate material and the temperature, on the behavior of FinFET device. The investigation was carried out on a Germanium-based FinFET device. The working device showed improvements on the current drivability in terms of high ON current (I_{ON}), less leakage current (I_{OFF}), and a high value of I_{ON}/I_{OFF} ratio and has reasonable control on short channel effects. Moreover, the analysis carried out reveals that a high work-function gate material with optimum temperature shows good electrostatic behavior.

One way to reduce the impact generated by WFF variations on full adders (FAs) is the replacement of internal inverters by Schmitt Triggers [76]. Four FAs were analyzed in nominal voltage and near-threshold regime at the layout level using the 7-nm FinFET node from ASAP7. In general, the Schmitt Trigger technique presented considerable robustness improvements over all full adders. The power robustness variability using the Schmitt Trigger technique can be up to 37.3% and 66.6% better than traditional architectures operating at the nominal and near-threshold regime, respectively. The main disadvantage is the area penalty highlighting the need for new design techniques to address variability at the layout level.

The evaluation of process variability and SET masking on a set of complex logic gates considering different transistor topologies is explored in [18] using the 7-nm FinFET electrical model from ASAP7. A comparison is made between complex logic gates in their traditional versions and a multi-level of basic logic gates that implement the same function using only NAND2, only NOR2, and NAND2/NOR2/INV cells. The functions were converted using De Morgan's theorem. Results show that although complex cells present better timing and power results, multi-level circuits are up to 28% less sensible to radiation faults and about 40% more stable under process variability.

According to the best results encountered in the previous work, a new study was done at the layout level using the ASAP7 technology [19]. Seven logic cells were designed using the complex logic gate and only NAND2 gates (providing a multi-level cell design). At nominal conditions, the complex gate topology presents the best results, but under the effects of transient faults or process variability, multi-level arrangements are the best option. Despite the area impact, NAND2 topology mitigates at least 50% of the impact on delay due to process variability effects reaching, on average, more than 85% of improvement compared to complex gates. Moreover, NAND2 topology improves over 45% on average the fault coverage evaluation from SET effects for these layouts.

Schmitt Triggers are promising circuits for variability effects mitigation and enhancement of noise immunity being widely applied on critical applications with reliability constraints. In [77], Schmitt Triggers were evaluated over multiple scenarios considering several levels of process variability, supply voltages, transistor sizing, and clock frequencies, prioritizing better energy consumption and the attenuation of process variability effects using the ASAP7 PDK. The hysteresis intervals showed attractive advantages of up to 10.8% and 25.3% when a higher number of fins and supply voltages were tested, respectively, bringing noise immunity improvements. It could be observed up to 16% and 44.7% maximum increase and decrease in the clock frequency, respectively, with differences between variability impact in the layouts, rising alongside the supply voltage values. The data set in this paper can provide relevant information for VLSI designers and the design of low-power applications that need to manage the process variability impact.

3.3 Radiation-Induced Soft Errors

Another key reliability concern at advanced technology nodes is the susceptibility to natural radiation environments. Initially, the radiation effects in electronic systems were only considered relevant in military, avionic, or spatial designs. However, with microelectronics advancement and low supply voltages, transient faults can occur even at sea level and may bring critical consequences [38].

The natural radiation can be divided into spatial and atmospheric environments. The three main spatial radiation sources are the solar wind, galactic cosmic rays, and belt radiations [13], as shown in Fig. 3.11. Solar radiation depends on sun activity. Typically, in a period of high solar activity, few neutrons are detected. Still, in a period of lower solar activity, the Earth's magnetic field traps particles that can be absorbed by the atmosphere. The cosmic radiation arises from stellar flares, supernova explosions, and other cosmic activities, consisting mainly of protons. Finally, the Van Allen belts are the space areas closest to the Earth with many protons and electrons. These protons are especially dangerous for spacecraft following the low Earth orbit (LEO).

The Earth is protected by a magnetic field that acts as a radioactive filter, blocking a large quantity of radiation from space (solar flares, solar winds, and cosmic rays). However, high energetic particles arising from cosmic radiation are not trapped by this filter, and they can interact with the atmosphere via direct ionization or by nuclear reactions. The nuclear reactions produce every kind of secondary radiation

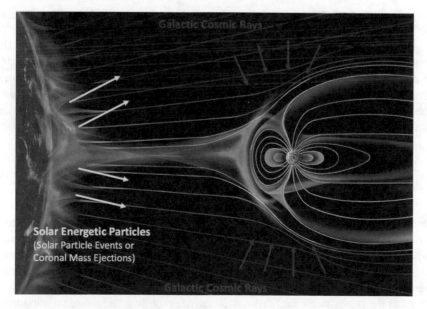

Fig. 3.11 Spatial radiation environment [30]

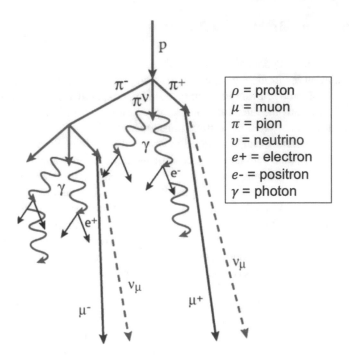

Fig. 3.12 Nuclear cascade particle reactions [74]

[13]. The result of the cosmic-ray shower is a set of energetic particles such as protons, electrons, neutrons, heavy ions, muons, and pions, as shown in Fig. 3.12. The type of secondary radiation and the intensity depend on the altitude, the geomagnetic latitude, and the Sun's activity. However, only 5% of protons and 1% of electrons and neutrons reach the Earth's surface at ground level from the total particles generated in the cascade effect [108]. This happens due to the loss of energy generated by the successive collisions and the life span of these particles. Currently, muons are the energetic particles most numerous at ground level [50].

3.3.1   Radiation Effects on Devices

Three categories of radiation effects affect the integrated circuits: displacement damage (DD), total ionizing dose (TID), and single event effects (SEEs). Displacement damage refers to the dislodging of atoms of the material crystalline structure due to non-ionizing energy loss (NIEL) of the incident particles. NIEL depends on the material being irradiated, the particle type, and the particle energy [109]. DD degrades the minority carrier lifetime, the carrier mobility, and the net doping level due to the introduction of new energy levels in the semiconductor band gap.

TID is a cumulative effect due to long-term silicon exposure to radiation, and it can permanently damage a circuit. The total accumulated dose depends on orbit altitude, orientation, and time. TID is measured concerning radiation absorbed dose (rad). The trapped charges in the STI oxide and its interface with silicon affect the electrical characteristics [117]. Irradiated circuits for the long term can cause threshold voltage shifts, increased leakage currents, timing changes, mobility degradation, and circuit functionality loss. Planar technologies needed hardening by design (HBD) techniques like enclosed layout transistor (ELT), or guard rings to become the integrated circuits almost free from the TID effects [41]. However, the 3D structure and the oxides used in the FinFET manufacturing process are favorable to attenuate TID's impact.

Single event effects occur due to the interaction of energetic particles with the silicon coming from space and atmospheric radiations. For older technologies, a transient pulse only happens if the collected charge (Q_{COLL}) exceeds the critical charge (Q_{CRIT}) of the nodes. The critical charge is defined as the minimal quantification of charge required to induce a SEE on a given node. However, nanometer technologies increase the proximity of devices, such that a single hit can diffuse the charge to the adjacent nodes introducing the concept of charge sharing [114].

The charge deposited by a single ionizing particle can produce a wide range of effects classified as destructive and non-destructive, as shown in Fig. 3.13. Destructive effects cause permanent and irreversible functional damages [103]. The most known destructive effects are Single Event Latchup (SEL) where, exclusively in CMOS devices, a low resistance path is created between the power supply and ground rails, Single Event Burnout (SEB) when the particle reaches the source

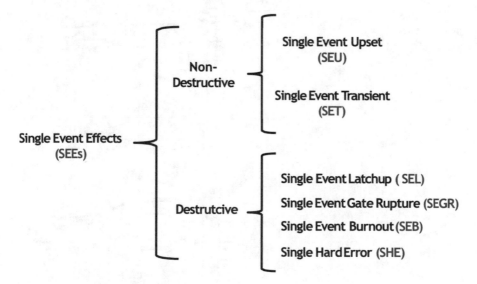

Fig. 3.13 Classification of single event effects [107]

region of the transistor, creating a conductive path between the source and the drain, Single Event Gate Rupture (SEGR) where the gate dielectric isolating the gate and channel regions fails, and Single Hard Errors (SHEs) when the deposition of large charges affects the state transitions of the devices.

Non-destructive effects, also named soft errors, induce a temporary deviation where the data are corrupted for a short time interval [87]. The most well-known non-destructive effects are Single Event Upset (SEU) when an energetic particle hits a sequential circuit, such as SRAM, latches or flip-flops, causing the change of the stored bit and Single Event Transient (SET) when the particle strikes a combinational circuit, such as basic gates or full adders, generating a transient pulse that may or may not be captured by a memory element. The interaction of radiation with the matter of the devices (mainly Si, SiO_2, Al, or Cu) is usually quantified by the linear energy transfer, which is a measure of the average ionizing energy deposited by a particle per unit path length [100].

Figure 3.14 exemplifies the non-destructive effects: (a) SET and (b) SEU. First, a SET occurs in a sensitive node of a NOR2 logic gate, and it generates a pulse at the stroke node. This pulse was propagated, and it reached the sequential logic to the right, which stored the incorrect value "0." Memory cells have two stable states,

Fig. 3.14 (**a**) Single event transient and (**b**) single event upset on a circuit

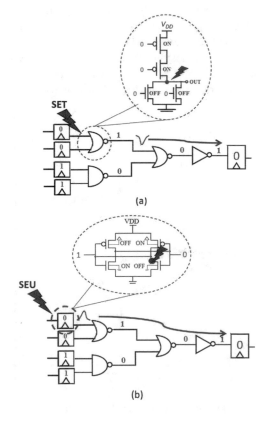

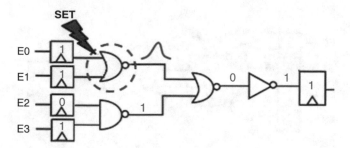

Fig. 3.14 Logical masking in a combinational circuit

one representing a stored value "0" and the other representing a stored value "1." In each state, there are two transistors in on-state and two transistors in off-state. In the second case, an energetic particle hits the sequential circuit in one sensitive node. So, a bit-flip happens, and it affects the rest of the circuit because the incorrect value is also captured by the sequential element on the right.

There are some cases where a transient pulse will be masked, and the sequential elements will not capture it. In this case, the fault will not lead to errors or failures visible to the user. Moreover, the circuit keeps a correct value in the output because the faults are masked still in origin. There are three kinds of masking observed in logic blocks: logical masking, electrical masking, and latch-window masking [66, 106]. The logical masking happens when a particle affects a portion of the circuit, but the hit node is not relevant to determine the final output. In this way, the output can be determined only by inputs not affected by radiation effects. For example, the first input of a NAND2 logic gate in Fig. 3.14 is "0," and then, the second input is not essential because the final result will always be "1." So, if a particle impacts one of the inputs, the error will not be seen in the final output. According to the truth table, the same happens with a NOR2 logic gate. If one input is equal to "1," the final result will always be "0."

The electrical masking happens when the fault impacts a circuit node, but the current pulse generated is attenuated through the combinational logic, disappearing before being stored by a forward latch. For example, in Fig. 3.15, the NOR2 logic gate has a SET in the first input, but the effect that it causes is mitigated when it is propagated until the output of an inverter. The fault reaches the forward latch, but the pulse has a small amplitude that is interpreted as a correct logical value, which, in this case, is equal to "0."

When a transient pulse cannot be masked logically or electrically, it propagates until it reaches a sequential circuit. Latch-window masking happens when a sequential logic does not capture the pulse. In Fig. 3.16, if the pulse at the NOR gate was not masked by one of the methods already presented, the memory element can mask it according to the latch-window. On the right of Fig. 3.16, it is shown a clock cycle with its latching window. If the SET is captured when a clock transition happened, a wrong value will be stored. Finally, the rate that SETs get latched as errors depends on the clock frequency and sequential circuits' topology.

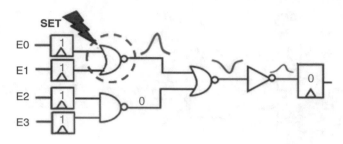

Fig. 3.15 Electrical masking in a combinational circuit

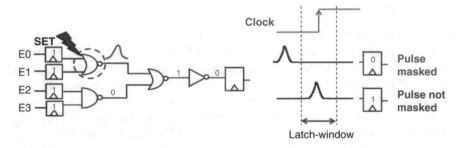

Fig. 3.16 Latch-window masking in a sequential circuit

In the scientific community, the two radiation effects most relevant are the total ionizing dose and single event effects [29]. Some years ago, TID was considered a major source of faults in integrated circuits. As technology has advanced, SEE gained more prominence and becomes a significant reliability concern for electronic systems in the space and ground levels. The TID effects were partially reduced due to the thinner oxides of the gate stack of modern deep-submicron processes [41]. However, the contribution of TID is now induced by charges trapped in STI interfaces. This point is enhanced in FinFET technology [7, 52, 59, 60].

On the other hand, transistors with shrinking geometry, higher speed, and logic density increase the SEE sensitivity. Moreover, as the supply voltage decreases, the charge stored at circuit nodes reduces according to Eq. 3.2, such that the critical charge can be larger than the collected charge more often. Consequently, the soft error susceptibility increases due to advanced technology nodes.

$$Q_{node} = C_{node} \times V_{DD} \tag{3.2}$$

The 3D structure of FinFETs presents attractive properties to control the increase of the radiation-induced soft errors compared to the bulk counterpart [39]. However, the change in device structure from planar to FinFET modifies the sensitive area and the charge collection mechanisms after an energetic particle hits the silicon [85]. In this way, the improvement of the reliability in sub-22nm technologies also requires the accurate understanding, predicting, and mitigating the single event effects on FinFET-based circuits.

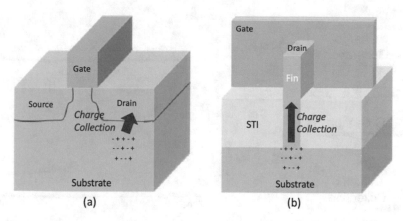

Fig. 3.17 Comparison of charge collection mechanism of (**a**) planar and (**b**) FinFET devices [65]

3.3.2   Charge Collection Mechanism

The collision of energetic particles causes a strong electric field perturbation due to the direct ionization, a primary mechanism of charge deposition caused by the incidence of alpha particles or heavy ions. As an energetic particle hits the silicon, it loses energy and forms a track of electron–hole pairs. If the ionization track transverses the depletion region, the electric field collects the carries generating a transient current pulse at the node. The charge generated by the impact of particles varies depending on the ion type, incident angle, and impact site.

In traditional planar devices, charges associated with the ion tracks colliding the silicon substrate are deposited in the drain directly and then diffuse to the drain, as shown in Fig. 3.17a. Otherwise, the thin fin region and the narrow connection to the substrate of the bulk FinFETs reduce the volume of silicon available for the charge collection compared with planar devices. Thus, a smaller amount of deposited charges can be expected to diffuse the FinFET drain, as illustrated in Fig. 3.17b [43]. For these reasons, the FinFET devices are considered less sensitive to soft errors.

The disturbance caused for the impact of energetic particles depends on the energy lost per unit track length, and it is known as linear energy transfer (LET). For every 3.6 eV (electron volts) of energy loss by the particle, one electron–hole pair is created in the silicon substrate [48]. The LET depends on the mass/energy of the particle and the material in which it is traveling. The highest LET values are obtained when more massive and energetic particles impact denser materials [14]. In this way, the pulse width is dependent on the particle energy, the charge stored at a node, and the charge collection in the affected junction.

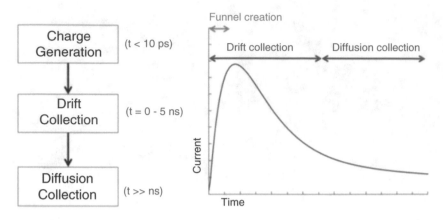

Fig. 3.18 Typical transient current waveform due to SEE [32]

After the silicon particle ionization, the process of charge collection proceeds through two mechanisms: drift and diffusion [78]. When the resultant ionization track traverses the depletion region, carriers are rapidly collected by the high electric field. This charge collection is known as drift. The crossing of particles through the depletion region is responsible for temporary deformation in a funnel shape. This effect is called funneling, and it causes an increase in the collected charge efficiency due to the rise of the depletion region area [14]. Finally, the diffusion process manages all the other carries generated besides the depletion layer. The typical transient current waveform resulting from the additionally collected charge induced by particle incidence can be seen in Fig. 3.18. The funnel creation and drift mechanism are high-speed processes. They are responsible for controlling the almost instantaneous rise of the transient current due to the deformation of the electric field of the junction. In the diffusion mechanism, a longer time is needed to collect the charge and, then, the transient pulse has a slower fall time.

Charge deposition mechanism proposed in [73] is widely used to form a current source whose behavior is modeled as a double exponential. The modeling of the transient current is given by Eqs. 3.3, 3.4, and 3.5, where Q_{COLL} is the collected charge due to a radiation particle strike, $\tau\alpha$ is the collection time constant of the junction, $\tau\beta$ is the ion track establishment time constant, and L is the charge collection depth that decreases with the technology scaling. In bulk silicon, a typical charge collection depth for a heavy ion is approximately $2\,\mu m$. For every $1\,MeV\,cm^2/mg$, an ionizing particle deposits about $10.8\,fC$ of electron–hole pairs along each micron of its track [68].

For sub-22nm technologies, especially for LETs higher than 10 MeV, the typical transient current waveform tends to suffer some modifications presenting a behavior similar to "plateau" [99]. Double exponential current sources still are considered the most reasonable first-order estimate, and it is widely adopted as a base model for SEE analysis [122]. However, this approach is extremely limited because of the empiric method, which, in the best case, dramatically overestimates the worst case of a single injection of SET. To be able to consider layout, voltage, and temperature

dependence of SET shapes, more realistic models have been developed such as
MRED from Vanderbilt University, Accuro from Robust Chip, SEMM from IBM,
Tiara from STMicroelectronics, and MUSCA SEP3 from ONERA. The details of
MUSCA SEP3 will be presented in the next subsection and Chap. 5.

$$I_P(t) = I_0 \times \left(e^{-\frac{t}{\tau_\alpha}} - e^{-\frac{t}{\tau_\beta}} \right) \tag{3.3}$$

$$I_0 = Q_{COLL}/(\tau_\alpha - \tau_\beta) \tag{3.4}$$

$$Q_{COLL} = 10.8 f \times L \times LET \tag{3.5}$$

3.3.3   Related Works

This subsection summarizes the main related works about radiation effects in
FinFET technologies available in the literature. A theoretical overview of the several
ways that FinFET circuits can be affected due to radiation was discussed in [100].
In general, the sensitivity of technology to SEE increases as device dimensions
decrease. The susceptibility of the SOI and bulk FinFETs to total ionizing dose
and single event effects was compared in [6]. Similar research was done in [96]
reporting the radiation experiments in UTBB FDSOI 28-nm for the first time.
In general, bulk FinFETs tend to collect more charge than SOI FinFETs due to
substrate considerations.

The fault models for FinFET circuits were discussed in [67] including fin stuck-
on, stuck-open, and gate oxide short. One single defect may affect multiple gates
that are correlated due to the configuration of FinFET. The effects of the Single
Event Latchup in bulk FinFETs and planar technologies were investigated by
[34]. The small fin structure seems to be harmful to latch-up immunity due to
the increased parasitic resistance and the reduced guard ring efficiency. However,
FinFET circuits with a higher supply voltage have a higher risk of latch-up effects.

The FinFET structure modifies the charge collection mechanism on a device.
A charge collection mechanism for FinFETs using through-wafer two-photon
absorption and ion beam experiments was analyzed in [39]. In [8], the behavior
of FinFET devices under radiation effects was investigated, and the SET pulse
modeling at advanced technology nodes. In [75], the behavior of FinFET devices
under radiation effects and the SET pulse modeling at advanced technology nodes
were investigated.

In [101], the radiation-induced soft error rates (SERs) of memory and logic
devices designed in the 22-nm Trigate technology were reported. A comparison
with the 32-nm planar devices showed an SER reduction of 1.5×–4× for cosmic
radiation. An overview of the impact of terrestrial radiation on soft error sensitivity

considering bulk, FDSOI, and FinFET technologies was done in [50]. According to the results, the muon is the main particle to provoke SER in sub-22nm technologies.

The MUSCA SEP3 was developed by [51]. This platform is dedicated to SEE prediction based on a Monte Carlo method, which allows a full simulation from the radiation environment definition to the soft error in the integrated circuit. The analysis confirms that MUSCA SEP3 agrees very well with TCAD simulations and SEU/SET irradiation testing while reducing the computational effort by several magnitude orders.

In planar technologies, a set of layout techniques was proposed to attenuate the radiation-induced soft errors. Although the FinFET structure is different from the planar one, some approaches can be tested to mitigate transient faults in FinFETs. A layout approach via pulse quenching was investigated by [10]. One extra drain diffusion is inserted into a cell output for this technique to intentionally promote charge sharing between transistors and quench the voltage pulse on the output. TCAD simulations demonstrated a reduction of 60% and 70% in the sensitive area and the pulse width, respectively.

A set of n-well contact schemes was investigated to verify the influence on the pulse width of SETs in [4]. Results showed a strong relationship between the SET pulse width and the percentage of the n-well area contacted, contradicting the theory that the cross section of a logic gate is based entirely on transistors' drain regions. A multi-finger layout technique for N-hit SET mitigation is discussed in [25] for the standard cell design. TCAD simulations with heavy ion experiments show that SET pulse widths are efficiently reduced with the multi-finger approach. Moreover, the method presented an area penalty acceptable, and the performance of circuits remains almost unchanged. Multi-finger is preferred over the traditional arrangement in the radiation-hardened integrated circuit design. Similarly, according to [27], a source isolation technique is used for the P-hit SET mitigation.

Another layout technique for SET mitigation based on dummy transistors was proposed in [26]. The approach calls for adding an off-state idle PMOS and NMOS transistor to the circuit connected to the sensitive region. Cell performance that uses the proposed layout almost does not change and generates a smaller area penalty.

The behavior of FinFET circuits under the total ionizing dose effects was explored in [24, 52, 60] and [123]. The TID response of bulk FinFETs is investigated under geometric variations [24]. Transistors with more extended channels (L_G) degrade less than those with shorter channels. On the other hand, devices with large fin pitch degrade more than those with narrow fin pitch. The TID-induced degradation increases with the decrease of the fin thickness (T_{SI}). TID radiation effects on 14-nm bulk and SOI FinFET technologies were analyzed by [52]. The replacement of MOSFET by FinFET device generates an increase in TID sensitivity due to the trapped charge in the STI oxide. Moreover, irradiation resulted in significant changes in the threshold voltage for SOI devices and significant deviations in the IOFF current for bulk FinFETs.

In [60], a near-threshold operation is presented as a methodology for reducing the increases in leakage currents caused by radiation. Results indicate devices with high channel stop doping as the most robust response to TID, allowing stable operation of ring oscillators and the SRAM bit-cell with a little shift in critical operating characteristics. The TID response was evaluated with strained Ge pMOS FinFETs varying the fin length, fin thickness, and gate length in [123]. Modest threshold voltage shifts, small transconductance degradation, and minimal changes in I_{ON}/I_{OFF} ratios are observed. In comparison with planar Ge pMOS, the improvements in Ge pMOS FinFETs happen due to the material quality, reductions in the STI thickness in areas of relevance to transistor operation, and better gate control.

Hardening techniques to attenuate the effects of radiation-induced soft errors were investigated in [21, 80] and [5]. A novel design style that reduces the impact of radiation-induced single event transient on logic circuits was presented in [21]. It enhances the robustness in noisy environments by strengthening the sensitive nodes using a technique similar to feedback. The results were compared with other techniques for hardening radiation at the transistor level using the 7-nm FinFET technology. Strengthening presents the best immunity in a noisy environment.

A charge steering-based latch hardening technique with significant SEU hardness was proposed in [80]. TCAD simulations showed the effect of charge steering in 16-nm bulk FinFET devices for reducing the amount of collected charge at the critical sensitive nodes considering alpha, proton, and heavy ions. The trade-offs regarding the area, performance, and power penalties are much lower compared to the other hardening approaches already proposed in the literature.

A set of techniques to mask SET and SEU using spatial, temporal, and hybrid redundancy was investigated by [5]. The performance and energy impact of each method are quantified at the near-threshold operation. A comparison between 45-nm planar and 16-nm FinFET technologies was made to investigate the effects of technology scaling. Temporal redundancy provides higher energy saving, but it requires consideration of the SET pulse duration.

The variability along with SET in FinFETs was explored in [9, 36, 102] and [37]. The soft error evaluation of logic gates in FinFET technologies considering the prediction tool MUSCA SEP3 coupled with electrical simulations of the gate was presented in [9]. First, a comparison of this tool with the TCAS mixed-mode simulation for an ion with LET equal to $5\,\mathrm{MeV\,cm^2\,mg^{-1}}$ was made, obtaining a good agreement between them as shown in Fig. 3.19. Only a small divergence can be observed in the NOR2 logic gate when the output voltage of NET1 and NET2 comes back to their initial logic state.

Two soft error reduction techniques were explored in NAND2 and NOR2 FinFET logic gates: the supply voltage variation and the drive strength. As discussed previously in this chapter, the higher supply voltage diminishes the SET susceptibility of both logic gates. Using a larger channel width in FinFETs, i.e., higher drive strength, the SET sensitivity of both logic gates decreases, as shown in Fig. 3.20. The NAND2 cell is immune to SETs induced by atmospheric neutrons when the X4 drive strength is used.

The SER of memory and logic devices manufactured in a 14-nm FinFET technology was measured in [102]. Results showed that SER is dominated by high-energy neutron-induced upset rates (77%), while thermal neutron and alpha particle contribute around 16% and 7%, respectively. Moreover, there is an SER reduction

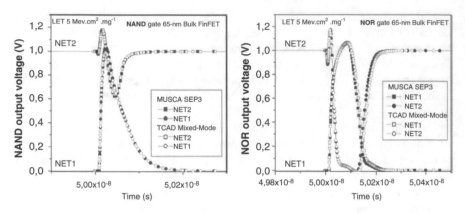

Fig. 3.19 Comparison of the transient response of 65-nm NAND2 and NOR2 gates using MUSCA SEP3 and TCAS mixed mode [9]

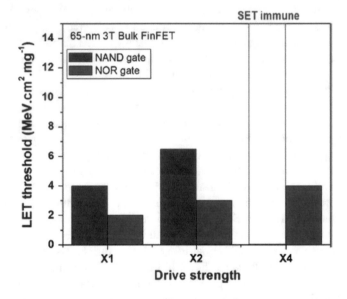

Fig. 3.20 LET threshold for NAND2 and NOR2 cells for different drive strength [9]

when the 14-nm node is adopted instead of the 22-nm technology. This reduction happens due to the fin dimensions, little scaling, and a decrease in charge collection efficiency per fin.

A comparative analysis of two majority voters based on NOR and NAND gates designed in the 7-nm FinFET technology to estimate the SER was explored by [36] at the layout level. In general, the NOR voter is less sensitive to soft errors than the NAND voter as it provides a lower soft error rate. However, NOR voters presented a larger SET pulse width. At nominal supply voltage, no event has been observed for alpha and atmospheric environment.

In [37], different XOR topologies under radiation effects were implemented using two multigate devices: double-gate FinFET and Trigate. Trigate-based circuits were demonstrated to be more robust than FinFET with an improvement percentage from 6.2% up to 12.6% in the LET threshold. Furthermore, voltage deviation can reduce the threshold LET up to 20.8%, increasing the fault susceptibility of the analyzed circuits.

The SEU susceptibility in FinFETs in sequential circuits was verified in [58, 81, 85, 86, 115, 116] and [124]. A detailed analysis of radiation-induced soft errors of SRAMs designed in SOI FinFET technology was presented by [58]. The authors considered the effects of supply voltage and process variations on the soft error rate of SRAM memory arrays. They conclude that SER is higher for lower supply voltages, MBU/SEU ratio is relatively higher for alpha radiation than that for protons, and neglecting the impact of process variation leads to an under-estimation of SER.

The SEU cross section over an extensive supply voltage range for D flip-flops designed in the 16-nm FinFET technology was evaluated in [81]. This work also considers a comparison between planar and FinFET devices concerning SEU sensitivity. The cross section increases with a reduction in bias for low-LET particles as alpha particles and low-energy protons. Results reinforce that SEU rates are influenced by the supply voltage and the operating environment.

Characterization of the soft error rate through the alpha irradiations considering combinational cells, SRAM, and flip-flops manufactured in the 14-nm FinFET technology was presented by [116]. The main factor contributing to the increase of SEU is the charge collection on NMOS transistors due to low-LET incidence. Design schemes for low power have little impact on the SER. An extension of this work was published in [115] where the 10-nm FinFET technology was evaluated.

A comparison of SEU trends among 16-nm bulk FinFET, 20-nm bulk planar, and 28-nm bulk planar was made in [85] considering D flip-flop. For LETs lower than $10\,\mathrm{MeV\,cm^2/mg}$, 16-nm FinFET flip-flops presented a considerably smaller SEU cross section than the planar technologies. However, the cross section of the 16-nm FinFET flip-flop for high-LET particles is very similar to planar nodes analyzed, as shown in Fig. 3.21. The SET pulse width is reduced in FinFET technology for low as well as high LETs. Moreover, the SET pulse width decreases when the supply voltage is increased. FinFET technology has a lower critical charge than that of planar nodes such that 3D simulations demonstrated a more considerable difference between them than experiments after the fabrication process.

Later, in [86], the authors investigated the FinFET structural effects on the SE cross section. Results showed that an ion strike direct at the fin produces an observable SET. In contrast, an ion strike between two fins results in a minimum voltage perturbation according to Fig. 3.22. For the planar technology, independently of the hit location, a SET pulse was observed at the inverter output for low-LET particles. For strikes by energetic particles with high LET, no dependence was observed to either device technologies, as shown in Fig. 3.23.

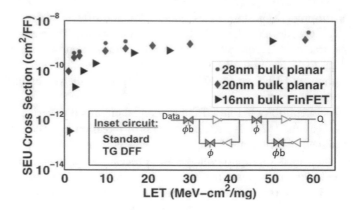

Fig. 3.21 SEU cross section versus LET for FinFET and planar technologies [85]

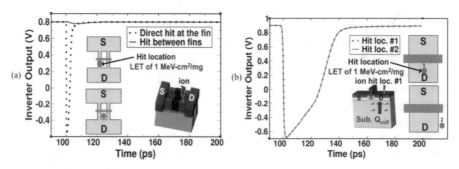

Fig. 3.22 Impact of transistor structure on low LET for (**a**) 16-nm FinFET and (**b**) 28-nm planar technologies using 3D TCAD simulations [86]

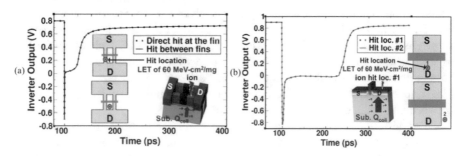

Fig. 3.23 Impact of transistor structure on high LET for (**a**) 16-nm FinFET and (**b**) 28-nm planar technologies using 3D TCAD simulations [86]

References

1. Abu-Rahma, M.H., Anis, M.: A statistical design-oriented delay variation model accounting for within-die variations. IEEE Trans. Comput. Aided Design Integr. Circuits Syst. **27**(11), 1983–1995 (2008)
2. Agarwal, S., Pandey, R.K., Johnson, J.B., Dixit, A., Bajaj, M., Furkay, S.S., Oldiges, P.J., Murali, K.V.R.M.: Ab initio study of metal grain orientation-dependent work function and its impact on FinFET variability. IEEE Trans. Electron Devices **60**(9), 2728–2733 (2013)
3. Aguiar, Y.Q., Zimpeck, A.L., Meinhardt, C., Reis, R.A.L.: Temperature dependence and ZTC bias point evaluation of sub 20nm bulk multigate devices. In: 2017 24th IEEE International Conference on Electronics, Circuits and Systems (ICECS), pp. 270–273 (2017)
4. Ahlbin, J.R., Atkinson, N.M., Gadlage, M.J., Gaspard, N.J., Bhuva, B.L., Loveless, T.D., Zhang, E.X., Chen, L., Massengill, L.W.: Influence of n-well contact area on the pulse width of single-event transients. IEEE Trans. Nuclear Sci. **58**(6), 2585–2590 (2011)
5. Alghareb, F.S., Ashraf, R.A., Alzahrani, A., DeMara, R.F.: Energy and delay tradeoffs of soft-error masking for 16-nm FinFET logic paths: survey and impact of process variation in the near-threshold region. IEEE Trans. Circuits Syst. II: Express Briefs **64**(6), 695–699 (2017)
6. Alles, M.L., Schrimpf, R.D., Reed, R.A., Massengill, L.W., Weller, R.A., Mendenhall, M.H., Ball, D.R., Warren, K.M., Loveless, T.D., Kauppila, J.S., Sierawski, B.D.: Radiation hardness of FDSOI and FinFET technologies. In: IEEE 2011 International SOI Conference, pp. 1–2 (2011)
7. Artola, L., Chiarella, T., Nuns, T., Cussac, G., Mitard, J.: Foundry dependence of total ionizing dose effects of FinFET transistor in 14-nm technological node. In: IEEE Nuclear Space Radiation Effects Conference (NSREC) (2020)
8. Artola, L., Gaillardin, M., Hubert, G., Raine, M., Paillet, P.: Modeling single event transients in advanced devices and ICS. IEEE Trans. Nuclear Sci. **62**(4), 1528–1539 (2015)
9. Artola, L., Hubert, G., Alioto, M.: Comparative soft error evaluation of layout cells in FinFET technology. Microelectro. Reliab. **54**(9), 2300–2305 (2014). SI: ESREF 2014
10. Atkinson, N.M., Witulski, A.F., Holman, W.T., Ahlbin, J.R., Bhuva, B.L., Massengill, L.W.: Layout technique for single-event transient mitigation via pulse quenching. IEEE Trans. Nuclear Sci. **58**(3), 885–890 (2011)
11. Avci, M., Babac, M.Y., Yildirim, T.: Neural network based MOSFET channel length and width decision method for analogue integrated circuits. Int. J. Electro. **92**(5), 281–293 (2005)
12. Bakshi, V.: EUV Lithography. SPIE Press Monograph. Society of Photo Optical. SPIE Press, Bellingham (2018)
13. Barth, J.L., Dyer, C.S., Stassinopoulos, E.G.: Space, atmospheric, and terrestrial radiation environments. IEEE Trans. Nuclear Sci. **50**(3), 466–482 (2003)
14. Baumann, R.C.: Radiation-induced soft errors in advanced semiconductor technologies. IEEE Trans. Device Mat. Reliab. **5**(3), 305–316 (2005)
15. Borkar, S.: Designing reliable systems from unreliable components: the challenges of transistor variability and degradation. IEEE Micro **25**(6), 10–16 (2005)
16. Borkar, S.: Design perspectives on 22nm CMOS and beyond. In: 2009 46th ACM/IEEE Design Automation Conference, pp. 93–94 (2009)
17. Brain, R.: 14nm Technology Leadership (2017). https://newsroom.intel.com/newsroom/wp-content/uploads/sites/11/2017/03/Ruth-Brain-2017-Manufacturing.pdf
18. Brendler, L.H., Zimpeck, A.L., Meinhardt, C., Reis, R.: Evaluating the impact of process variability and radiation effects on different transistor arrangements. In: 2018 IFIP/IEEE International Conference on Very Large Scale Integration (VLSI-SoC), pp. 71–76 (2018)
19. Brendler, L.H., Zimpeck, A.L., Meinhardt, C., Reis, R.: Exploring multi-level design to mitigate variability and radiation effects on 7nm FinFET logic cells. In: 2018 25th IEEE International Conference on Electronics, Circuits and Systems (ICECS), pp. 581–584 (2018)

20. Cao, Y., Clark, L.T.: Mapping statistical process variations toward circuit performance variability: an analytical modeling approach. In: Proceedings of the 42nd Design Automation Conference, 2005, pp. 658–663 (2005)
21. Calomarde, A., Amat, E., Moll, F., Vigara, J., Rubio, A.: Set and noise fault tolerant circuit design techniques: Application to 7nm FinFET. Microelectron. Reliab. **54**(4), 738–745 (2014)
22. Cathignol, A., Cheng, B., Chanemougame, D., Brown, A.R., Rochereau, K., Ghibaudo, G., Asenov, A.: Quantitative evaluation of statistical variability sources in a 45-nm technological node LP N-MOsfet. IEEE Electron Device Lett. **29**(6), 609–611 (2008)
23. Chang, H., Sapatnekar, S.: Circuit timing and leakage power analysis under process variations. Thesis (PhD) – University of Minnesota (2006)
24. Chatterjee, I., Zhang, E.X., Bhuva, B.L., Reed, R.A., Alles, M.L., Mahatme, N.N., Ball, D.R., Schrimpf, R.D., Fleetwood, D.M., Linten, D., Simôen, E., Mitard, J., Claeys, C.: Geometry dependence of total-dose effects in bulk FinFETs. IEEE Trans. Nuclear Sci. **61**(6), 2951–2958 (2014)
25. Chen, J., Chen, S., He, Y., Chi, Y., Qin, J., Liang, B., Liu, B.: Novel layout technique for n-hit single-event transient mitigation via source-extension. IEEE Trans. Nuclear Sci. **59**(6), 2859–2866 (2012)
26. Chen, J., Chen, S., He, Y., Qin, J., Liang, B., Liu, B., Huang, P.: Novel layout technique for single-event transient mitigation using dummy transistor. IEEE Trans. Device Mater. Reliab. **13**(1), 177–184 (2013)
27. Chen, J., Chen, S., Liang, B., Liu, B.: Simulation study of the layout technique for p-hit single-event transient mitigation via the source isolation. IEEE Trans. Device Mater. Reliab. **12**(2), 501–509 (2012)
28. Clark, L.T., Vashishtha, V., Shifren, L., Gujja, A., Sinha, S., Cline, B., Ramamurthy, C., Yeric, G.: ASAP7: A 7-nm FinFET predictive process design kit. Microelectron. J. **53**, 105–115 (2016)
29. Clemens, M.A.: Energy deposition mechanisms for proton-and-neutron-induced single event upsets. Thesis (PhD) – Vanderbilt University (2012)
30. Chancellor, J.C., Scott, G.B., Sutton, J.P.: Space radiation: The number one risk to astronaut health beyond low earth orbit. Life (Basel) **4**(3), 491–510 (2014)
31. Choi, J.H., Murthy, J., Roy, K.: The effect of process variation on device temperature in FinFET circuits. In: 2007 IEEE/ACM International Conference on Computer-Aided Design, pp. 747–751 (2007)
32. Cummings, D.J.: Enhancements in CMOS device simulation for single-event effects. Thesis (PhD) – University of Florida (2010)
33. Dadgour, H., Vivek De, Banerjee, K.: Statistical modeling of metal-gate work-function variability in emerging device technologies and implications for circuit design. In: 2008 IEEE/ACM International Conference on Computer-Aided Design, pp. 270–277 (2008)
34. Dai, C., Chen, S., Linten, D., Scholz, M., Hellings, G., Boschke, R., Karp, J., Hart, M., Groeseneken, G., Ker, M., Mocuta, A., Horiguchi, N.: Latchup in bulk FinFET technology. In: 2017 IEEE International Reliability Physics Symposium (IRPS), pp. EL–1.1–EL–1.3 (2017)
35. Das, R., Baishya, S.: Investigation of work function and temperature of germanium FinFETs. In: 2017 International Conference on Electron Devices and Solid-State Circuits (EDSSC), pp. 1–2 (2017)
36. de Aguiar, Y., Artola, L., Hubert, G., Meinhardt, C., Kastensmidt, F., Reis, R.: Evaluation of radiation-induced soft error in majority voters designed in 7nm FinFET technology. Microelectron. Reliab. **76–77**, 660–664 (2017)
37. de Aguiar, Y.Q., Meinhardt, C., Reis, R.A.L.: Radiation sensitivity of XOR topologies in multi-gate technologies under voltage variability. In: 2017 IEEE 8th Latin American Symposium on Circuits Systems (LASCAS), pp. 1–4 (2017)
38. Dodd, P.E., Shaneyfelt, M.R., Schwank, J.R., Felix, J.A.: Current and future challenges in radiation effects on CMOS electronics. IEEE Trans. Nuclear Sci. **57**(4), 1747–1763 (2010)

39. El-Mamouni, F., Zhang, E.X., Pate, N.D., Hooten, N., Schrimpf, R.D., Reed, R.A., Galloway, K.F., McMorrow, D., Warner, J., Simoen, E., Claeys, C., Griffoni, A., Linten, D., Vizkelethy, G.: Laser- and heavy ion-induced charge collection in bulk FinFETs. IEEE Trans. Nuclear Sci. **58**(6), 2563–2569 (2011)

40. Endo, K., Matsukawa, T., Ishikawa, Y., Liu, Y.X., O'uchi, S., Sakamoto, K., Tsukada, J., Yamauchi, H., Masahara, M.: Variation analysis of tin FinFETs. In: 2009 International Semiconductor Device Research Symposium, pp. 1–2 (2009)

41. Faccio, F.: Design Hardening Methodologies for ASICs, pp. 143–160. Springer, Dordrecht (2007)

42. Fan, M., Wu, Y., Hu, V.P., Su, P., Chuang, C.: Investigation of cell stability and write ability of FinFET subthreshold SRAM using analytical SNM model. IEEE Trans. Electron Devices **57**(6), 1375–1381 (2010)

43. Fang, Y., Oates, A.S.: Neutron-induced charge collection simulation of bulk FinFET SRAMs compared with conventional planar SRAMs. IEEE Trans. Device Mater. Reliab. **11**(4), 551–554 (2011)

44. Ghai, D., Mohanty, S.P., Kougianos, E.: Design of parasitic and process-variation aware nano-CMOS RF circuits: A VCO case study. IEEE Trans. Very Large Scale Integr. Syst. **17**(9), 1339–1342 (2009)

45. GSS: Case Study: Statistical Variability in an Example 22nm FinFET (2010). http://www.goldstandardsimulations.com/GSS_22nm_FinFET_case_study.pdf

46. Gupta, P., Papadopoulou, E.: Yield Analysis and Optimization. CiteSeerX (2011)

47. Harish, B.P., Bhat, N., Patil, M.B.: On a generalized framework for modeling the effects of process variations on circuit delay performance using response surface methodology. IEEE Trans. Comput. Aided Design Integr. Circuits Syst. **26**(3), 606–614 (2007)

48. Hartmann, F.: Silicon Detectors (2009). http://lxmi.mi.infn.it/ISAPP/editionsold/2009karlsruhe/www.kceta.kit.edu/downloads/Hartmann_Si-Detectors.pdf

49. Henderson, C.L.: Failure analysis techniques for a 3d world. Microelectron. Reliab. **53**(9), 1171–1178 (2013). European Symposium on Reliability of Electron Devices, Failure Physics and Analysis

50. Hubert, G., Artola, L., Regis, D.: Impact of scaling on the soft error sensitivity of bulk, FDSOI and FinFET technologies due to atmospheric radiation. Integration **50**, 39–47 (2015)

51. Hubert, G., Duzellier, S., Inguimbert, C., Boatella-Polo, C., Bezerra, F., Ecoffet, R.: Operational SER calculations on the SAC-C orbit using the multi-scales single event phenomena predictive platform (MUSCA SEP3). IEEE Trans. Nuclear Sci. **56**(6), 3032–3042 (2009)

52. Hughes, H., McMarr, P., Alles, M., Zhang, E., Arutt, C., Doris, B., Liu, D., Southwick, R., Oldiges, P.: Total ionizing dose radiation effects on 14 nm FinFET and SOI UTBB technologies. In: 2015 IEEE Radiation Effects Data Workshop (REDW), pp. 1–6 (2015)

53. Islam, A., Akram, M.W., Hasan, M.: Variability immune FinFET-based full adder design in subthreshold region. In: 2011 International Conference on Devices and Communications (ICDeCom), pp. 1–5 (2011)

54. Janakiraman, V., Bharadwaj, A., Visvanathan, V.: Voltage and temperature aware statistical leakage analysis framework using artificial neural networks. IEEE Trans. Comput. Aided Design Integr. Circuits Syst. **29**(7), 1056–1069 (2010)

55. Jiang, X., Guo, S., Wang, R., Wang, X., Cheng, B., Asenov, A., Huang, R.: A device-level characterization approach to quantify the impacts of different random variation sources in FinFET technology. IEEE Electron Device Lett. **37**(8), 962–965 (2016)

56. Karapetyan, S., Kleeberger, V., Schlichtmann, U.: FinFET-based product performance: modeling and evaluation of standard cells in FinFET technologies. Microelectron. Reliab. **61**, 30–34 (2016). SI: ICMAT 2015

57. Khalid, U., Mastrandrea, A., Olivieri, M.: Effect of NBTI/PBTI aging and process variations on write failures in mosfet and FinFET flip-flops. Microelectron. Reliab. **55**(12, Part B), 2614–2626 (2015)

58. Kiamehr, S., Osiecki, T., Tahoori, M., Nassif, S.: Radiation-induced soft error analysis of srams in SOI FinFET technology: A device to circuit approach. In: 2014 51st ACM/EDAC/IEEE Design Automation Conference (DAC), pp. 1–6 (2014)

59. King, M.P., Massey, G., Silva, A., Cannon, E., Shaneyfelt, M.R., Loveless, T., Ballast, J., Cabanas-Holmen, M., DiGregorio, S., Rice, W.C., Draper, B.L., Oldgies, P., Rodbell, K.: TID-Induced Leakage and Drive Characteristics of Planar 22-nm PDSOI and 14-nm Bulk and Quasi-SOI FinFET Devices. Sandia National Lab.(SNL-NM), Albuquerque (2018)

60. King, M.P., Wu, X., Eller, M., Samavedam, S., Shaneyfelt, M.R., Silva, A.I., Draper, B.L., Rice, W.C., Meisenheimer, T.L., Felix, J.A., Zhang, E.X., Haeffner, T.D., Ball, D.R., Shetler, K.J., Alles, M.L., Kauppila, J.S., Massengill, L.W.: Analysis of TID process, geometry, and bias condition dependence in 14-nm FinFETs and implications for RF and SRAM performance. IEEE Trans. Nuclear Sci. **64**(1), 285–292 (2017)

61. Kleeberger, V.B., Graeb, H., Schlichtmann, U.: Predicting future product performance: Modeling and evaluation of standard cells in FinFET technologies. In: 2013 50th ACM/EDAC/IEEE Design Automation Conference (DAC), pp. 1–6 (2013)

62. Klein, R.: Overview of process variability. In: Proceedings of the ISSCC Microprocessor Forum F6: Transistor Variability Nanometer-Scale Technology, pp. A1–A24 (2008)

63. Kumar, U.S., Rao, V.R.: Thermal performance of nano-scale SOI and bulk FinFETs. In: 2016 15th IEEE Intersociety Conference on Thermal and Thermomechanical Phenomena in Electronic Systems (ITherm), pp. 1566–1571 (2016)

64. Lee, Y., Shin, C.: Impact of equivalent oxide thickness on threshold voltage variation induced by work-function variation in multigate devices. IEEE Trans. Electron Devices **64**(5), 2452–2456 (2017)

65. Lee, S., Kim, I., Ha, S., Yu, C., Noh, J., Pae, S., Park, J.: Radiation-induced soft error rate analyses for 14 nm FinFET SRAM devices. In: 2015 IEEE International Reliability Physics Symposium, pp. 4B.1.1–4B.1.4 (2015)

66. Liden, P., Dahlgren, P., Johansson, R., Karlsson, J.: On latching probability of particle induced transients in combinational networks. In: Proceedings of IEEE 24th International Symposium on Fault- Tolerant Computing, pp. 340–349 (1994)

67. Liu, Y., Xu, Q.: On modeling faults in FinFET logic circuits. In: 2012 IEEE International Test Conference, pp. 1–9 (2012)

68. Mavis, D.G., Eaton, P.H.: Soft error rate mitigation techniques for modern microcircuits. In: 2002 IEEE International Reliability Physics Symposium. Proceedings. 40th Annual (Cat. No.02CH37320), pp. 216–225 (2002)

69. McLellan, P.: FinFET Custom Design (2014). https://semiwiki.com/semiconductor-manufacturers/tsmc/3327-FinFET-custom-design/

70. Meinhardt, C.: Variabilidade em FinFETs. Thesis (Doutorado em Ciência da Computação) – Instituto de Informática - UFRGS (2014)

71. Meinhardt, C., Zimpeck, A., Reis, R.: Predictive evaluation of electrical characteristics of sub-22nm FinFET technologies under device geometry variations. Microelectron. Reliab. **54**(9), 2319–2324 (2014). SI: ESREF 2014

72. Meinhardt, C., Zimpeck, A.L., Reis, R.: Impact of gate workfunction fluctuation on FinFET standard cells. In: 2014 21st IEEE International Conference on Electronics, Circuits and Systems (ICECS), pp. 574–577 (2014)

73. Messenger, G.C.: Collection of charge on junction nodes from ion tracks. IEEE Trans. Nuclear Sci. **29**(6), 2024–2031 (1982)

74. Mèszáros, P., Razzaque, S., Wang, X.Y.: Cosmic Ray Physics (2015). http://personal.psu.edu/nnp/cr.html

75. Monga, U., Choi, J., Jeon, J., Kwon, U., Lee, K., Choo, S., Uemura, T., Lee, S., Pae, S.: Charge-collection modeling for SER simulation in FinFETs. In: 2016 International Conference on Simulation of Semiconductor Processes and Devices (SISPAD), pp. 295–298 (2016)

76. Moraes, L., Zimpeck, A., Meinhardt, C., Reis, R.: Evaluation of variability using schmitt trigger on full adders layout. Microelectron. Reliab. **88–90**, 116–121 (2018). 29th European Symposium on Reliability of Electron Devices, Failure Physics and Analysis (ESREF 2018)

77. Moraes, L.B., Zimpeck, A.L., Meinhardt, C., Reis, R.: Minimum energy FinFET schmitt trigger design considering process variability. In: 2019 IFIP/IEEE 27th International Conference on Very Large Scale Integration (VLSI-SoC), pp. 88–93 (2019)

78. Munteanu, D., Autran, J.: Modeling and simulation of single-event effects in digital devices and ICS. IEEE Trans. Nuclear Sci. **55**(4), 1854–1878 (2008)

79. Mutlu, A.A., Rahman, M.: Statistical methods for the estimation of process variation effects on circuit operation. IEEE Trans. Electron. Packaging Manuf. **28**(4), 364–375 (2005)

80. Narasimham, B., Chandrasekharan, K., Wang, J.K., Ni, K., Bhuva, B.L., Schrimpf, R.D.: Charge-steering latch design in 16 nm FinFET technology for improved soft error hardness. IEEE Trans. Nuclear Sci. **64**(1), 353–358 (2017)

81. Narasimham, B., Hatami, S., Anvar, A., Harris, D.M., Lin, A., Wang, J.K., Chatterjee, I., Ni, K., Bhuva, B.L., Schrimpf, R.D., Reed, R.A., McCurdy, M.W.: Bias dependence of single-event upsets in 16 nm FinFET d-flip-flops. IEEE Trans. Nuclear Sci. **62**(6), 2578–2584 (2015)

82. Nassif, S.R.: Process variability at the 65nm node and beyond. In: 2008 IEEE Custom Integrated Circuits Conference, pp. 1–8 (2008)

83. Nawaz, S.M., Mallik, A.: Effects of device scaling on the performance of junctionless FinFETs due to gate-metal work function variability and random dopant fluctuations. IEEE Electron Device Lett. **37**(8), 958–961 (2016)

84. Neuberger, G., Wirth, G., Reis, R.: Protecting Chips Against Hold Time Violations Due to Variability. Springer, Dordrecht (2014)

85. Nsengiyumva, P., Ball, D.R., Kauppila, J.S., Tam, N., McCurdy, M., Holman, W.T., Alles, M.L., Bhuva, B.L., Massengill, L.W.: A comparison of the SEU response of planar and FinFET D flip-flops at advanced technology nodes. IEEE Trans. Nuclear Sci. **63**(1), 266–272 (2016)

86. Nsengiyumva, P., Massengill, L.W., Alles, M.L., Bhuva, B.L., Ball, D.R., Kauppila, J.S., Haeffner, T.D., Holman, W.T., Reed, R.A.: Analysis of bulk FinFET structural effects on single-event cross sections. IEEE Trans. Nuclear Sci. **64**(1), 441–448 (2017)

87. O'Bryan, M.: Radiation Effects and Analysis: Single Event Effects (2000). https://radhome.gsfc.nasa.gov/radhome/see.htm

88. Orshansky, M., Nassif, S., Boning, D.: Design for Manufacturability and Statistical Design. Springer, Berlin (2008)

89. Othman, N.A.F., Hatta, S.F.W.M., Soin, N.: Impacts of fin width scaling on the electrical characteristics of 10-nm FinFET at different metal gate work function. In: 2017 IEEE Regional Symposium on Micro and Nanoelectronics (RSM), pp. 256–259 (2017)

90. Posser, G., Belomo, J., Meinhardt, C., Reis, R.: Performance improvement with dedicated transistor sizing for mosfet and FinFET devices. In: 2014 IEEE Computer Society Annual Symposium on VLSI, pp. 418–423 (2014)

91. Posser, G., Sapatnekar, S.S., Reis, R.: Electromigration Inside Logic Cells: Modeling, Analyzing and Mitigating Signal Electromigration in NanoCMOS. Springer International Publishing, New York (2017)

92. Rasouli, S.H., Endo, K., Banerjee, K.: Variability analysis of FinFET-based devices and circuits considering electrical confinement and width quantization. In: 2009 IEEE/ACM International Conference on Computer-Aided Design - Digest of Technical Papers, pp. 505–512 (2009)

93. Rathore, R.S., Rana, A.K., Sharma, R.: Threshold voltage variability induced by statistical parameters fluctuations in nanoscale bulk and SOI FinFETs. In: 2017 4th International Conference on Signal Processing, Computing and Control (ISPCC), pp. 377–380 (2017)

94. Rathore, R.S., Sharma, R., Rana, A.K.: Impact of work function fluctuations on threshold voltage variability in a nanoscale FinFETs. In: 2016 IEEE International Symposium on Nanoelectronic and Information Systems (iNIS), pp. 261–263 (2016)

95. Rieger, M.L.: Communication theory in optical lithography. J. Micro/Nanolith. MEMS MOEMS **11**(1), 1–11 (2012)

96. Roche, P., Autran, J., Gasiot, G., Munteanu, D.: Technology downscaling worsening radiation effects in bulk: SOI to the rescue. In: 2013 IEEE International Electron Devices Meeting, pp. 31.1.1–31.1.4 (2013)

97. Saha, S.K.: Modeling process variability in scaled CMOS technology. IEEE Design Test Comput. **27**(2), 8–16 (2010)
98. Saha, R., Bhowmick, B., Baishya, S.: Si and Ge step-FinFETs: work function variability, optimization and electrical parameters. Superlattices Microstruct. **107**, 5–16 (2017)
99. Sayil, S.: Soft Error Mechanisms, Modeling and Mitigation, pp. 1–105. Springer International Publishing, New York (2016)
100. Schrimpf, R.D., Alles, M.A., Mamouni, F.E., Fleetwood, D.M., Weller, R.A., Reed, R.A.: Soft errors in advanced CMOS technologies. In: 2012 IEEE 11th International Conference on Solid-State and Integrated Circuit Technology, pp. 1–4 (2012)
101. Seifert, N., Gill, B., Jahinuzzaman, S., Basile, J., Ambrose, V., Shi, Q., Allmon, R., Bramnik, A.: Soft error susceptibilities of 22 nm tri-gate devices. IEEE Trans. Nuclear Sci. **59**(6), 2666–2673 (2012)
102. Seifert, N., Jahinuzzaman, S., Velamala, J., Ascazubi, R., Patel, N., Gill, B., Basile, J., Hicks, J.: Soft error rate improvements in 14-nm technology featuring second-generation 3d tri-gate transistors. IEEE Trans. Nuclear Sci. **62**(6), 2570–2577 (2015)
103. Sexton, F.W.: Destructive single-event effects in semiconductor devices and ICS. IEEE Trans. Nuclear Sci. **50**(3), 603–621 (2003)
104. Sherlekar, D.D.: Design considerations for regular fabrics. In: Proceedings of the 2004 International Symposium on Physical Design, ISPD '04, p. 97–102. Association for Computing Machinery (2004)
105. Shin, C., Kim, J., Kim, G., Lee, H., Shin, C., Kim, J., Cho, B.J., Yu, H.: Random dopant fluctuation-induced threshold voltage variation-immune Ge FinFET with metal–interlayer–semiconductor source/drain. IEEE Trans. Electron Devices **63**(11), 4167–4172 (2016)
106. Shivakumar, P., Kistler, M., Keckler, S.W., Burger, D., Alvisi, L.: Modeling the effect of technology trends on the soft error rate of combinational logic. In: Proceedings International Conference on Dependable Systems and Networks, pp. 389–398 (2002)
107. Siegle, F., Vladimirova, T., Ilstad, J., Emam, O.: Mitigation of radiation effects in SRAM-based FPGAs for space applications. ACM Comput. Surv. **47**(2) (2015)
108. Simionovski, A.: Sensor de corrente transiente para detecção do set com célula de memória dinâmica. Dissertation (Master) – Engenharia Elétrica, UFRGS (2012)
109. Srour, J.R., Palko, J.W.: Displacement damage effects in irradiated semiconductor devices. IEEE Trans. Nuclear Sci. **60**(3), 1740–1766 (2013)
110. Swahn, B., Hassoun, S., Alam, S., Botha, D., Vidyarthi, A.: Thermal Analysis of FinFETs and Its Application to Gate Sizing (2005). http://www.tauworkshop.com/PREVIOUS_/05_Slides/tau05-3.1.pdf
111. Swahn, B., Hassoun, S.: Gate sizing: FinFETs vs 32nm bulk MOSFETs. In: 2006 43rd ACM/IEEE Design Automation Conference, pp. 528–531 (2006)
112. Thakker, R.A., Sathe, C., Baghini, M.S., Patil, M.B.: A table-based approach to study the impact of process variations on FinFET circuit performance. IEEE Trans. Comput. Aided Design Integr. Circuits Syst. **29**(4), 627–631 (2010)
113. Tomida, K., Hiraga, K., Dehan, M., Hellings, G., Jang, D., Miyaguhi, K., Chiarella, T., Kim, M., Mocuta, A., Horiguchi, N., Mercha, A., Verkest, D., Thean, A.: Impact of fin shape variability on device performance towards 10nm node. In: 2015 International Conference on IC Design Technology (ICICDT), pp. 1–4 (2015)
114. Toure, G., Hubert, G., Castellani-Coulie, K., Duzellier, S., Portal, J.: Simulation of single and multi-node collection: Impact on SEU occurrence in nanometric sram cells. IEEE Trans. Nuclear Sci. **58**(3), 862–869 (2011)
115. Uemura, T., Lee, S., Kim, G., Pae, S.: Investigation of logic soft error and scaling effect in 10 nm FinFET technology. In: 2017 IEEE International Reliability Physics Symposium (IRPS), pp. 2E-3.1–2E-3.4 (2017)
116. Uemura, T., Lee, S., Park, J., Pae, S., Lee, H.: Investigation of logic circuit soft error rate (SER) in 14nm FinFET technology. In: 2016 IEEE International Reliability Physics Symposium (IRPS), pp. 3B-4-1–3B-4-4 (2016)

117. Velazco, R., Fouillat, P., Reis, R. (Eds.): Radiation Effects on Embedded Systems, pp. 1–269. Springer Nature, London (2007)
118. Wang, X., Brown, A.R., Binjie Cheng, Asenov, A.: Statistical variability and reliability in nanoscale FinFETs. In: 2011 International Electron Devices Meeting, pp. 5.4.1–5.4.4 (2011)
119. Wang, X., Cheng, B., Brown, A.R., Millar, C., Alexander, C., Reid, D., Kuang, J.B., Nassif, S., Asenov, A.: Unified compact modelling strategies for process and statistical variability in 14-nm node DG FinFETs. In: 2013 International Conference on Simulation of Semiconductor Processes and Devices (SISPAD), pp. 139–142 (2013)
120. Wang, Y., Cotofana, S.D., Fang, L.: Statistical reliability analysis of NBTI impact on FinFET SRAMs and mitigation technique using independent-gate devices. In: 2012 IEEE/ACM International Symposium on Nanoscale Architectures (NANOARCH), pp. 109–115 (2012)
121. Wei, X., Zhu, H., Zhang, Y., Zhao, C.: Bulk FinFETs with body spacers for improving fin height variation. Solid-State Electron. 122, 45–51 (2016)
122. Wrobel, F., Dilillo, L., Touboul, A.D., Pouget, V., Saigné, F.: Determining realistic parameters for the double exponential law that models transient current pulses. IEEE Trans. Nuclear Sci. 61(4), 1813–1818 (2014)
123. Zhang, E.X., Fleetwood, D.M., Hachtel, J.A., Liang, C., Reed, R.A., Alles, M.L., Schrimpf, R.D., Linten, D., Mitard, J., Chisholm, M.F., Pantelides, S.T.: Total ionizing dose effects on strained Ge pMOS FinFETs on bulk Si. IEEE Trans. Nuclear Sci. 64(1), 226–232 (2017)
124. Zhang, H., Jiang, H., Assis, T.R., Ball, D.R., Narasimham, B., Anvar, A., Massengill, L.W., Bhuva, B.L.: Angular effects of heavy-ion strikes on single-event upset response of flip-flop designs in 16-nm bulk FinFET technology. IEEE Trans. Nuclear Sci. 64(1), 491–496 (2017)
125. Zhang, H., Jiang, H., Assis, T.R., Ball, D.R., Ni, K., Kauppila, J.S., Schrimpf, R.D., Massengill, L.W., Bhuva, B.L., Narasimham, B., Hatami, S., Anvar, A., Lin, A., Wang, J.K.: Temperature dependence of soft-error rates for FF designs in 20-nm bulk planar and 16-nm bulk FinFET technologies. In: 2016 IEEE International Reliability Physics Symposium (IRPS), pp. 5C–3–1–5C–3–5 (2016)
126. Zimpeck, A., Meinhardt, C., Reis, R.: Impact of PVT variability on 20nm FinFET standard cells. Microelectron. Reliab. 55(9), 1379–1383 (2015). Proceedings of the 26th European Symposium on Reliability of Electron Devices, Failure Physics and Analysis
127. Zimpeck, A.L., Aguiar, Y., Meinhardt, C., Reis, R.: Geometric variability impact on 7nm trigate combinational cells. In: 2016 IEEE International Conference on Electronics, Circuits and Systems (ICECS), pp. 9–12 (2016)
128. Zimpeck, A.L., Aguiar, Y., Meinhardt, C., Reis, R.: Robustness ofs sub-22nm multigate devices against physical variability. In: 2017 IEEE International Symposium on Circuits and Systems (ISCAS), pp. 1–4 (2017)
129. Zimpeck, A.L., Meinhardt, C., Posser, G., Reis, R.: Process variability in FinFET standard cells with different transistor sizing techniques. In: 2015 IEEE International Conference on Electronics, Circuits, and Systems (ICECS), pp. 121–124 (2015)
130. Zimpeck, A.L., Meinhardt, C., Posser, G., Reis, R.: FinFET cells with different transistor sizing techniques against PVT variations. In: 2016 IEEE International Symposium on Circuits and Systems (ISCAS), pp. 45–48 (2016)
131. Zimpeck, A.L., Meinhardt, C., Reis, R.: Evaluating the impact of environment and physical variability on the i_{ON} current of 20nm FinFET devices. In: 2014 24th International Workshop on Power and Timing Modeling, Optimization and Simulation (PATMOS), pp. 1–8 (2014)

Chapter 4
Circuit-Level Mitigation Approaches

Several techniques can be applied in different abstraction levels for enhancing the reliability of integrated circuits. Usually, TCAD simulations are used to measure the effectiveness of mitigation techniques based on the usage of other devices, materials, or doping profiles. Although this abstraction level presents very accurate results, it demands more considerable computational time for VLSI designs.

Then, one alternative is to investigate circuit-level approaches to achieve more robust solutions. Design adjustments can be related to the insertion of components or filtering elements, the exploration of different transistor arrangements, gate upsizing, transistor folding, hardware redundancy, increase of the capacitance of the most susceptible nodes, and the use of multi-level design instead of complex cells. This chapter focuses on explaining four circuit-level mitigation approaches.

4.1 Techniques Overview

Circuit-level methods are promising alternatives to achieve more robust solutions, with fewer penalties, and smaller cost of implementation. This book considers four approaches based on circuit-level to provide reliability improvement: transistor reordering, decoupling cells, Schmitt Trigger, and sleep transistor. They were chosen due to the potential to improve noise stability, helping to decrease process variations. Moreover, capacitive methods increase the critical charge to induce a soft error, enhancing the FinFET cell robustness.

All these methods were previously evaluated in the literature, focusing on improving other reliability challenges or using planar technologies. However, as FinFET nodes have different properties, the potential of them needed to be better investigated. Figure 4.1 shows the generic representation of all techniques in the circuit design. The schematic of the transistor reordering method is omitted because the modifications happen inside the pull-up or pull-down networks. The insertion of

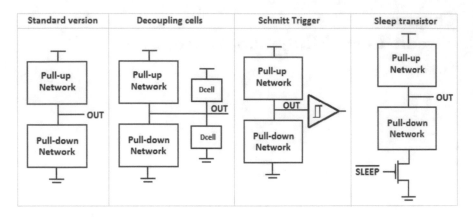

Fig. 4.1 Generic representation of the circuit-level mitigation approaches

decoupling cells or Schmitt Trigger in the gate output does not modify the traditional schematic of logic cells. On the other hand, the addition of a sleep transistor changes only the pull-down network, with a new transistor connected to the ground rail.

The main characteristics, advantages, drawbacks, and the details about the implementation are presented in the next subsections, adopting the AOI21 logic cell as an example. The same methodology can be adopted for any other logic cell.

4.2 Transistor Reordering

The optimization of the transistor arrangements is a method typically used to design faster circuits, to reduce the leakage currents, or to deal with BTI effects [2, 5, 6]. This technique aims to modify the transistor arrangements keeping the same logic function for all topologies. The possibilities can be obtained using different logic styles such as complementary CMOS, ratioed logic, domino logic, and pass-transistor logic, or transistor reordering.

The transistor reordering changes the electrical and physical characteristics of the logic cells, and consequently, the susceptibility to process variations and soft errors is also modified. Figure 4.2 shows two alternative topologies for the AOI21 cell that are logically equivalent. In the pull-up network, the serial transistor (input signal A) can be placed close or far to the cell output. As a general rule, AOI gates have the alternative transistor arrangements explored in the pull-up network, while the OAI gates explored different pull-down network options. Some logic gates, such as AOI221 and OAI221, can also investigate an intermediate place between the parallel associations to put the serial transistor.

The close topology is defined as the standard version in this work because it is the most used in the standard cell libraries. When the transistors of the complementary network have only parallel associations, as shown in Fig. 4.2, the rearrangement is

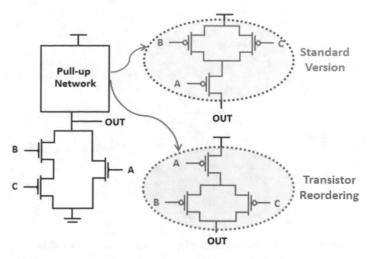

Fig. 4.2 Standard version of AOI21 cell and applying transistor reordering method

unnecessary because it does not influence the power consumption and performance results. The absence of area penalties is the main advantage of the transistor reordering technique.

4.3 Decoupling Cells

The addition of decoupling cells is a capacitive method commonly used in several industrial designs to ensure higher noise immunity on the supply rails and signal lines [8, 11]. Decoupling cells are connected in the gate output. They are composed of two transistors arranged in the cross-coupled mode, as shown in Fig. 4.3 for the AOI21 cell, where the transistor sizing of decoupling cells defines the total capacitance.

These decoupling cells increase the total capacitance in the output node, increasing the critical charge to produce a SET pulse and making this node less susceptible to the impact of energetic particles. In [1], this technique was used to filter SET pulses generated by low energy particles in a set of logic gates. They were designed using the Innovations for High-Performance (IHP) 130-nm bulk CMOS digital library with the fault injection through the double-exponential current at the SPICE level. Moreover, decoupling cells deliver current to the gates during the switching, improving the signal disturbances caused by process variations.

Two decoupling cells are required to ensure effective soft error hardening for both positive and negative polarities of the SET pulses. One decoupling cell is connected between the output and the supply rail, while the other is placed between the output and ground rail. As the insertion of decoupling cells is a capacitive method, larger

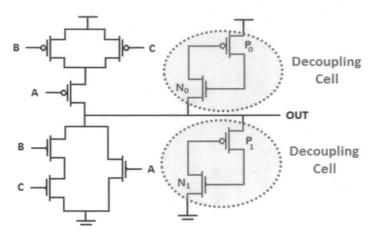

Fig. 4.3 Design of the AOI21 logic cell connecting decoupling cells in the output

decoupling cells contribute even more to the attenuation of process variability and radiation-induced soft errors. The disadvantage is the area and delay overheads due to four more transistors in the design. Otherwise, the advantage is that cross-coupled transistors do not impact power consumption as much.

4.4   Schmitt Trigger

Schmitt Triggers have an inherent hysteresis property commonly used to enhance signal stability and high noise immunity. This work explores a well-known topology of Schmitt Trigger where the main difference from the most common versions is the presence of P_F and N_F devices that are responsible for a feedback scheme, as shown in Fig. 4.4 [9]. For example, if the output is high, the N_F transistor is on, pulling the node X to a high potential. It forces the drain–source voltage of transistor N_1 to almost zero and its gate–source voltage into the negative region. This kind of topology reduces the leakage current in N_1 exponentially, increasing the I_{ON}-to-I_{OFF} current ratio, and minimizing the output degradation.

One of the process variability effects is a shift in the voltage transfer curve (VTC) due to the threshold voltage variation. The variability impact on VTC is reduced with a Schmitt Trigger due to the strong influence of the gate–source voltage of the inner transistors (N_1 and P_1) over its switching point. The replacement of traditional inverters by Schmitt Triggers on full adders shows to be an attractive alternative to mitigate the effects of process variations on planar technologies [7, 12] and also for a FinFET technology [10]. The main drawback of this technique is also the area and power overheads due to the addition of six more transistors in the circuit.

Fig. 4.4 Design of the
AOI21 logic cell adding a
Schmitt Trigger in the output

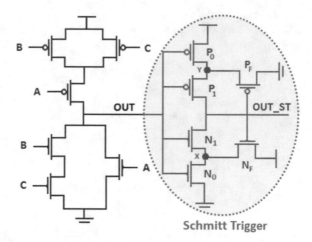

Schmitt Trigger

4.5 Sleep Transistors

The power-gating is one strategy widely employed in low-power designs to shut off
circuit blocks that are not in use, improving the overall power on a chip [4]. The
difference among the power-gating designs is the granularity of the blocks. This
work focuses on a fine-grained model where a sleep transistor is added to every cell.
However, for larger circuits, the block-grained style is more indicated to avoid the
area overhead. Figure 4.5 illustrates the AOI21 cell with a sleep transistor placed
between the pull-down network and the ground rail.

The sleep signal is used to control the "active" (sleep = 0) and "idle" (sleep = 1)
states of the transistor. When the sleep transistor is in active mode, it guarantees a
typical connection from the logic cell to the ground rail, acting as a supply voltage
regulator. The sleep transistor is turned off in the standby mode, disconnecting the
virtual ground (V_{GND}) from the physical ground. This behavior aims to reduce
leakage currents, transient faults, and NBTI effects.

Moreover, the addition of sleep transistors proved to be very efficient to mitigate
the impact of process variations in planar technologies [3]. However, two fundamental points must be considered for the sleep transistor technique to be successfully
applied: (1) the correct control of the sleep signal and (2) the adoption of proper
sizing. The main disadvantage of this technique is the performance degradation
when the sleep transistor is in the active mode, leading this path to become the
worst-case delay of logic cells. On the other hand, this technique imposes only one
extra transistor, introducing a small area overhead.

Fig. 4.5 Design of the
AOI21 logic cell using a sleep
transistor

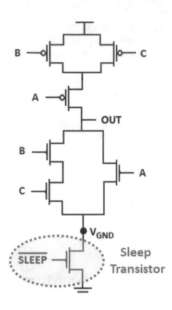

References

1. Andjelkovic, M., Babic, M., Li, Y., Schrape, O., Krstic, M., Kraemer, R.: Use of decoupling cells for mitigation of set effects in CMOS combinational gates. In: 2018 25th IEEE International Conference on Electronics, Circuits and Systems (ICECS), pp. 361–364 (2018)
2. Butzen, P.F., Bem, V.D., Reis, A.I., Ribas, R.P.: Transistor network restructuring against NBTI degradation. Microelectro. Reliab. **50**(9), 1298–1303 (2010). 21st European Symposium on the Reliability of Electron Devices, Failure Physics and Analysis
3. Calimera, A., Macii, A., Macii, E., Poncino, M.: Power-Gating for Leakage Control and Beyond, pp. 175–205. Springer, New York (2015)
4. Calimera, A., Macii, E., Poncino, M.: NBTI-aware power gating for concurrent leakage and aging optimization. In: Proceedings of the 2009 ACM/IEEE International Symposium on Low Power Electronics and Design, ISLPED '09, p. 127–132. Association for Computing Machinery, New York (2009)
5. Chun, J.W., Chen, C.R.: Transistor and pin reordering for leakage reduction in CMOS circuits. Microelectron. J. **53**, 25–34 (2016)
6. da Silva, D.N., Reis, A.I., Ribas, R.P.: CMOS logic gate performance variability related to transistor network arrangements. Microelectron. Reliab. **49**(9), 977–981 (2009). 20th European Symposium on the Reliability of Electron Devices, Failure Physics and Analysis
7. Dokania, V., Islam, A.: Circuit level design technique to mitigate impact of process, voltage and temperature variations in complementary metal-oxide semiconductor full adder cells. IET Circuits Devices Syst. **9**(3), 204–212 (2015)
8. Evans, R.J., White, D.B., Pomerleau, R., Scougal, A.: Control of SSO noise through the use of on-chip decoupling cells. In: 2002 IEEE 11th Topical Meeting on Electrical Performance of Electronic Packaging, pp. 201–204 (2002)
9. Lotze, N., Manoli, Y.: A 62 mV 0.13 μm CMOS standard-cell-based design technique using schmitt-trigger logic. In: 2011 IEEE International Solid-State Circuits Conference, pp. 340–342 (2011)

10. Moraes, L., Zimpeck, A., Meinhardt, C., Reis, R.: Evaluation of variability using schmitt trigger on full adders layout. Microelectron. Reliab. **88–90**, 116–121 (2018). 29th European Symposium on Reliability of Electron Devices, Failure Physics and Analysis (ESREF 2018)
11. Su, H., Sapatnekar, S.S., Nassif, S.R.: Optimal decoupling capacitor sizing and placement for standard-cell layout designs. IEEE Trans. Comput. Aided Design Integr. Circuits Syst. **22**(4), 428–436 (2003)
12. Toledo, S.P., Zimpeck, A.L., Reis, R., Meinhardt, C.: Pros and cons of schmitt trigger inverters to mitigate PVT variability on full adders. In: 2018 IEEE International Symposium on Circuits and Systems (ISCAS), pp. 1–5 (2018)

Chapter 5
Evaluation Methodology

The main goals of this work are to evaluate the impact of process variability and radiation-induced soft errors at the physical level in a set of FinFET logic cells besides investigating circuit-level approaches to mitigate the unwanted effects caused by them. This chapter presents all the methodological design flows to achieve the purposed objectives, highlighting all the crucial details of the experimental setup.

5.1 Design Flow

The set of basic and complex cells evaluated in this work are INV, NAND2, NAND3, NAND4, NOR2, NOR3, NOR4, AOI21, OAI21, AOI211, and OAI211. The complex cells are defined as ones that have both serial and parallel transistor networks. The AND–OR inverter (AOI) is the two-level logic function composed of one or more AND gates that precede a NOR gate. The complementary of AOI cells is the OR–AND inverter (OAI) such that a NAND gate follows the OR gates. This set of logic gates was chosen for representing the most common cells among the standard libraries. More detailed information about them can be seen in Table 5.1.

The design flow used in this work is presented in Fig. 5.1. First, this design flow was performed considering the standard version of cells for comparison purposes. After, the schematic of each cell was changed using the circuit-level approaches described in Chap. 4 to obtain more reliable circuits. According to most standard cell libraries, the standard version of complex cells considers the serial transistors close to the cell output. This work considers the same transistor sizing (three fins) for all transistors of the gates to avoid overly difficult routing, or poor density [10]. The transistor sizing (three to five fins) is only applied in the extra transistors imposed by decoupling cells, Schmitt Trigger, and sleep transistor techniques.

© The Author(s), under exclusive license to Springer Nature Switzerland AG 2021
A. Zimpeck et al., *Mitigating Process Variability and Soft Errors at Circuit-Level for FinFETs*, https://doi.org/10.1007/978-3-030-68368-9_5

Table 5.1 Information about
the FinFET logic cells

Gates	Number of inputs	Number of transistors	Area (nm^2)
INV	1	2	50.9
NAND2	2	4	67.8
NAND3	3	6	84.8
NAND4	4	8	101.7
NOR2	2	4	67.8
NOR3	3	6	84.8
NOR4	4	8	101.7
AOI21	3	6	84.8
AOI211	4	8	101.7
OAI21	3	6	84.8
OAI211	4	8	101.7

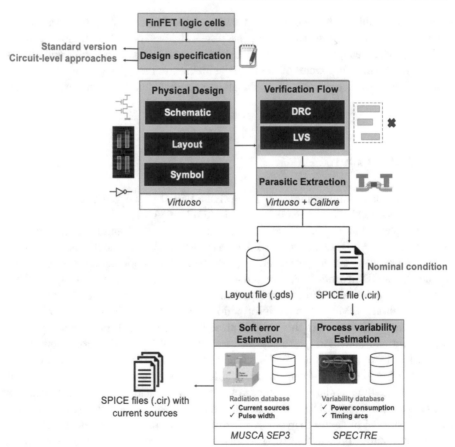

Fig. 5.1 Complete design flow

The logic cells pass by three elementary steps: physical design, verification flow, and parasitic extraction. In the physical design, the schematic, layout, and symbol of all logic cells were implemented using Cadence's Virtuoso tool. Since logic cells are used several times in the same integrated circuit, a cell library is a way to save time and avoid errors in the physical design. The height of logic cells was set as 7.5, 9, and 10.5 tracks of metal 2 (M2) when transistors with three, four, and five fins are used in the design, respectively. This work follows this design pattern to allow the future development of a cell library focused on reliability issues.

Each layout was submitted to the verification flow composed of a design rule check (DRC) and layout versus schematic (LVS) steps. DRC checks if the layout satisfies the design rules, such as width, pitch, spacing, area, overlap, and enclosure, required for the layers of a given technology. On the other hand, the LVS compares devices and connectivities presented in the schematic with those of the layout. In this work, DRC and LVS steps are based on the technology rules of the 7-nm FinFET PDK called ASAP7 developed by ASU in collaboration with ARM Ltd [7]. This PDK implements the shorted-gate model, where a TAP cell is used to connect the back-gate to the front-gate, ensuring proper functionality of transistors.

The layer information and the basic design rules of this PDK are summarized in Table 2.4 and Fig. 2.11. Finally, the parasitic wire resistances and capacitances (RC) are extracted from the layout. A new circuit netlist is generated such that each net has one subckt with the RC tree structure and the connections between the parasitic networks. Caliber tool from Mentor and Virtuoso tool were used to perform the verification flow and parasitic extraction steps. The geometric data stream (GDS) is a file generated by Virtuoso that represents all the geometric shapes of layout design in the binary format. This file can reconstruct all or part of a layout, transfer the layout between different tools, or create photomasks for the fabrication process.

A SPICE file was created with all details to perform electrical simulations such as input waves, supply voltage, all metrics to be measured (power, propagation delays, currents), technological model, and the new netlist coming from the parasitic extraction step. This work considers FinFETs in a bulk substrate for all experiments. For a more realistic assessment, all logic cells drive a fan-out 4 (FO4). Moreover, two inverters are connected to each input to represent a load. The SPICE file is simulated to verify the nominal conditions, i.e., the behavior of each cell without process variability or radiation effects. All the electrical simulations were carried out using SPECTRE from Cadence. The details about process variability insertion and radiation analysis will be discussed in the next subsections.

5.2 Process Variability Evaluation

Monte Carlo (MC) is the most common method used to model the probability of different outcomes that cannot easily be predicted due to the many random variables involved. According to [1], two thousand MC simulations is enough to obtain accurate results in the variability analysis. For these reasons, this work considers

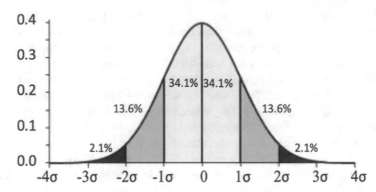

Fig. 5.2 Normal distribution curve [6]

two thousand MC simulations performed in SPECTRE from Cadence to estimate the behavior of FinFET logic cells under process variations. As presented in Chapter 3, work-function is the most impacted parameter by process variability in FinFET technologies. In this work, the WF is modeled as a Gaussian function, assuming 3-sigma (σ) deviation, which represents 99.7% of the normal distribution curve, as shown in Fig. 5.2.

All logic cells were evaluated using WFF levels varying from 1% to 5% due to the lack of information from the industry about the levels of WFF in the current FinFET technologies. These variations were adopted as a reference to the nominal values of the regular threshold voltage (RVT) model from ASAP7 at a typical configuration. The geometric parameters and doping information of this model can be seen in Table 2.2. The variability database provides a summary of statistical results from MC simulations such as the minimum (min) and maximum (max) values, mean (μ), and standard deviation (σ) for all timing arcs and also for power consumption.

The standard deviation quantifies the variation of a set of data from the nominal conditions. A low standard deviation indicates that the data is more close to the mean. Despite the data provided by the variability database for the delay and power metrics, this work also adopted two figures of merit to allow a more detailed comparison: (1) the normalized standard deviation (σ/μ) to indicate the sensitivity of logic cells to the WF fluctuations and (2) the delta relation (Δ) to specify how much the sensitivity to process variation changes when logic cells are designed using a circuit-level mitigation approach instead of the standard version.

The σ/μ relation was calculated for all timing arcs. From this, there are two ways to evaluate the delay variability. First, the higher σ/μ relation among all timing arcs is considered, and after, the σ/μ relation of the worst-case delay is used as a reference (the propagation delay with the largest mean). This book presents the results of delay variability considering both methodologies. The relative deviations in power and delay metrics are estimated by comparing the nominal values (without any variation) with the mean values of Monte Carlo simulations.

For example, the AOI21 cell was designed considering the standard version and adopting the four circuit-level techniques presented in Chap. 4. Each design has a different σ/μ relation for delay and power metrics. A design is pointed out as the best choice to mitigate the delay or power variability if it has the lowest value for the σ/μ relation. Otherwise, the delta relation compares the σ/μ relation of the standard version with each of those obtained using circuit-level methods. A technique is classified as favorable to power or delay variability mitigation if the delta relation has positive values. The same idea was followed to evaluate the other cells.

5.3   Soft Error Estimation

The SE susceptibility of FinFET logic cells was estimated using the MUSCA SEP3 tool developed by ONERA, the French Aerospace Lab [8, 9]. MUSCA SEP3 is a radiation event generator tool based on the Monte Carlo method, which models all stages since the strike of an energetic particle into the matter until the manifestation of a transient pulse on the circuits. The radiation particles available for analysis are the neutrons, protons, heavy ions, muons, and alpha particles. This tool takes into account the targeted radiation environment (space, avionics, ground), radiation features (LET, angle of incidence, type of energetic particle), dynamic transport and charge collection mechanisms, the bias voltage, the layout characteristics, the electrical circuit response, the STI oxide, and the details of the manufacturing process. Figure 5.3 summarizes the design flow executed by the MUSCA SEP3 tool.

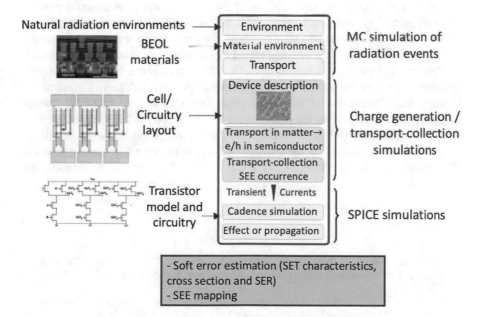

Fig. 5.3 MUSCA SEP3 prediction flow for FinFET technology nodes [2]

The collection charges of each transistor are calculated based on the layout of the device, which can be extracted from reverse engineering or using the FEOL report from the GDSII file. The modeling transport and the collection of free carriers in the silicon are performed using 3D analytical models, through the BEOL information, that adopt the following mechanisms: ambipolar diffusion, dynamic collection, multi-collection bipolar amplification to evaluate the charge sharing and pulse quenching phenomenon, recombination, and bias dependence [3, 5]. The impact of the temperature (down to 50K) is considered to all the physical and electrical models used for the transport and collection of charge in the semiconductors. The model implemented for the bipolar amplification depends on two aspects [4]. First, the model uses the equivalent access resistances of the multigate device to determine the triggering of the bipolar transistor. Also, the model considers the variability of the amplification of charge collection as a function of LET due to the FinFET technologies.

The SET database generated by the tool is very accurate since it considers all characteristics presented above. To each different setup, a new SET database is created. Once done, it is composed of a set of current sources to be injected in the sensitive nodes, i.e., the drain of transistors. The fault injection is performed automatically using a script along with SPECTRE from Cadence. After, the results are evaluated to determine the soft error susceptibility of the circuits.

This work explores the heavy ion irradiation at a normal angle of incidence, room temperature (27 °C), and with the supply voltage varying from the nominal value (0.7 V) down to the near-threshold regime (0.3 V). NOR2, NAND2, and AOI21 logic cells were evaluated under low LET, i.e., less than $15 \, \text{MeV} \, \text{cm}^2 \, \text{mg}^{-1}$, which corresponds to the representative secondary particles induced by neutrons or protons in avionics and ground applications. The SE susceptibility was also investigated under higher LETs (30 and $58 \, \text{MeV} \, \text{cm}^2 \, \text{mg}^{-1}$) representing the space environment.

The SET database was simulated for all input vectors to obtain a more accurate estimation. Moreover, the output of each cell is connected to a chain of four inverters, allowing the evaluation of propagation effects. A fault is accounted for if the voltage amplitude of the output node exceeds the gate threshold voltage ($V_{DD}/2$). This work adopts the cross section (σ_{cs}) as the central figure of merit to estimate the SE susceptibility of logic cells. This metric quantifies the probability of an energetic particle crossing the area of $1 \, \text{cm}^2$ and to produce a transient event. The relation between the SE susceptibility and the SET pulse width were also analyzed.

References

1. Alioto, M., Consoli, E., Palumbo, G.: Variations in nanometer CMOS flip-flops: Part i–impact of process variations on timing. IEEE Trans. Circuits Syst. I: Regular Papers **62**(8), 2035–2043 (2015)
2. Artola, L., Hubert, G., Alioto, M.: Comparative soft error evaluation of layout cells in FinFET technology. Microelectro. Reliab. **54**(9), 2300–2305 (2014). SI: ESREF 2014

3. Artola, L., Hubert, G., Duzellier, S., Bezerra, F.: Collected charge analysis for a new transient model by TCAD simulation in 90 nm technology. IEEE Trans. Nuclear Sci. **57**(4), 1869–1875 (2010). https://doi.org/10.1109/TNS.2010.2053944

4. Artola, L., Hubert, G., Schrimpf, R.D.: Modeling of radiation-induced single event transients in SOI FinFETs. In: 2013 IEEE International Reliability Physics Symposium (IRPS), pp. SE.1.1–SE.1.6 (2013). https://doi.org/10.1109/IRPS.2013.6532108

5. Artola, L., Hubert, G., Warren, K.M., Gaillardin, M., Schrimpf, R.D., Reed, R.A., Weller, R.A., Ahlbin, J.R., Paillet, P., Raine, M., Girard, S., Duzellier, S., Massengill, L.W., Bezerra, F.: SEU prediction from set modeling using multi-node collection in bulk transistors and SRAMs down to the 65 nm technology node. IEEE Trans. Nuclear Sci. **58**(3), 1338–1346 (2011). https://doi.org/10.1109/TNS.2011.2144622

6. Chandler, D.L.: How do you know when a new finding is significant? the sigma value can tell you – but watch out for dead fish. (2012). https://news.mit.edu/2012/explained-sigma-0209

7. Clark, L.T., Vashishtha, V., Shifren, L., Gujja, A., Sinha, S., Cline, B., Ramamurthy, C., Yeric, G.: ASAP7: A 7-nm FinFET predictive process design kit. Microelectron. J. **53**, 105–115 (2016)

8. Hubert, G., Artola, L.: Single-event transient modeling in a 65-nm bulk CMOS technology based on multi-physical approach and electrical simulations. IEEE Trans. Nuclear Sci. **60**(6), 4421–4429 (2013)

9. Hubert, G., Duzellier, S., Inguimbert, C., Boatella-Polo, C., Bezerra, F., Ecoffet, R.: Operational SER calculations on the SAC-C orbit using the multi-scales single event phenomena predictive platform (MUSCA SEP3). IEEE Trans. Nuclear Sci. **56**(6), 3032–3042 (2009)

10. Vashishtha, V., Vangala, M., Clark, L.T.: Asap7 predictive design kit development and cell design technology co-optimization: Invited paper. In: 2017 IEEE/ACM International Conference on Computer-Aided Design (ICCAD), pp. 992–998 (2017)

Chapter 6
Process Variability Mitigation

Process variability is a random deviation in the device structure, which causes an increase or decrease in typical design specifications. These deviations can affect the reliability of circuits because it can modify the I_{ON}/I_{OFF} currents, the power consumption, the performance, and the threshold voltage. First, this chapter presents an overview of the impact of LER and MGG variations on the I_{ON} and I_{OFF} currents of FinFET devices using a 7-nm technology. After, the behavior of FinFET logic cells in the standard version is evaluated, considering the nominal conditions (i.e., without process variations) and under work-function fluctuations. The standard version corresponds to the evaluation of logic cells without any circuit-level mitigation technique. Finally, this chapter evaluates the adoption of circuit-level approaches in the design of FinFET logic cells to improve the robustness of logic cells. This chapter also presents the mitigation tendency when different levels of WF fluctuations and transistor sizing were used.

The results presented in this chapter seek: (1) to reinforce that process variations can modify the behavior of logic cells significantly, and consequently, (2) to highlight the importance of proposing techniques able to mitigate the effects caused by this challenge. Moreover, the results of this chapter will be used as a reference point to estimate how much the adoption of circuit-level approaches in the design improves the robustness of logic cells.

6.1 Device Characterization

The influence of deviations in the gate length (L_G), fin height (H_{FIN}), fin width (W_{FIN}), and work-function (WF) due to LER and MGG variabilities was evaluated, focusing on the I_{ON}/I_{OFF} currents of PFET and NFET devices from ASAP7. Figure 6.1 illustrates the impact of process variations on the I_{ON} current of NFET devices with the minimum transistor sizing (1 fin). For the geometric parameters,

© The Author(s), under exclusive license to Springer Nature Switzerland AG 2021
A. Zimpeck et al., *Mitigating Process Variability and Soft Errors at Circuit-Level for FinFETs*, https://doi.org/10.1007/978-3-030-68368-9_6

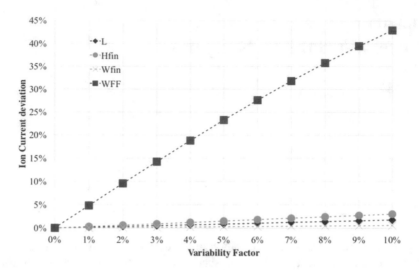

Fig. 6.1 Impact of geometric and WF variations on the I_{ON} current [1]

even considering 10% of deviation from nominal conditions, the impact on I_{ON} current is small, i.e., less than 5%. On the other hand, low levels of WF fluctuations already introduce at least 5% of deviation on the I_{ON} current. The impact of WF fluctuations on the I_{ON} current grows linearly with the increase in the levels of variation.

The density curves for the Monte Carlo simulations of each parameter, considering 10% of process variation, are shown in Fig. 6.2, for the I_{ON} current. It is possible to observe that geometric parameters present denser results due to the small deviations. PFET devices are less dense than NFET devices, such that the data is further to the mean. However, WF fluctuations show a significant deviation, with the impact on the I_{ON} current ranging from nano to micro-amperes.

Among the statistical variabilities on the geometric parameters, the deviations of H_{FIN} and L_G modify the I_{OFF} current up to 3% and 7%, respectively, as shown in Fig. 6.3. However, the roughness in the W_{FIN} may provoke more than 10% of deviation if the fabrication process introduces more than 5% of process variation. Figure 6.4 shows the influence of WF fluctuations on the I_{OFF} current for PFET and NFET devices. The WF fluctuations produce I_{OFF} currents drastically higher than the nominal behavior, mainly for PFET devices. Some of these highest values can be considered inaccurate because they are very similar to the I_{ON} currents. Regardless, the WF was already pointed out as the most impacted parameter by process variations in FinFET technologies.

The density curves for the I_{OFF} current under process variations are presented in Fig. 6.5 for both devices. Geometric parameters, such as gate length and fin width, demonstrate a reduction in the density for PFET devices, while the fin height becomes denser. However, NFET devices are more impacted by the fin

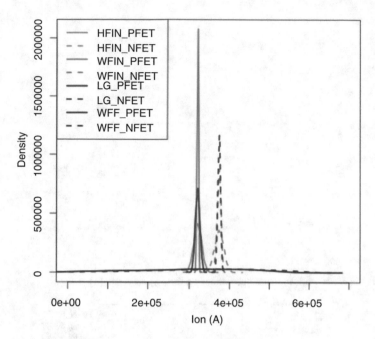

Fig. 6.2 Density curves of the I_{ON} current under process variations [1]

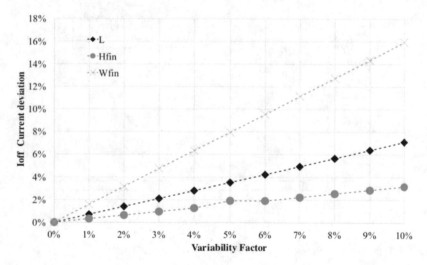

Fig. 6.3 Impact of geometric variations on the I_{OFF} current [1]

width variations, presenting a wider density curve. The WF fluctuations show large deviation mainly for PFET devices, even at small levels of variation.

The characterization of devices adopting the electrical model provided by ASAP7 is consistent with that obtained using other multigate technologies [2].

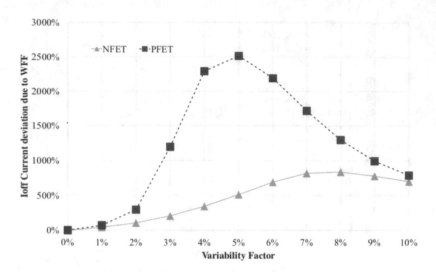

Fig. 6.4 Impact of work-function fluctuations on the I_{OFF} current [1]

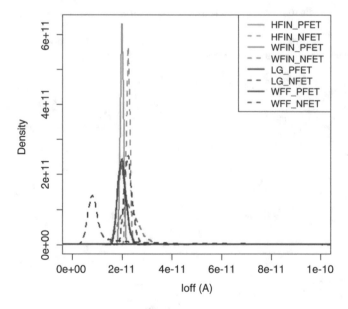

Fig. 6.5 Density curves for the I_{OFF} current under process variations

The work-function fluctuations dominate the impact on the I_{ON}/I_{OFF} currents of
FinFET devices. The increase in the number of fins is a way to protect the devices
against geometric variations, but this methodology does not work to attenuate the
WF fluctuation effects. For this reason, henceforward, this book always adopts the
work-function fluctuations for all process variability assessments.

6.2   Standard Design

The WF fluctuations on FinFET devices can generate power and delay deviations. These deviations are estimated using the normalized standard deviation (σ/μ relation). In this work, we adopt the terminology "power variability" and "delay variability" to indicate the deviations on power and delay due to WF fluctuations, respectively. Tables 6.1 and 6.2 show the typical values for the propagation delay (σ/μ relation of worst case) and power consumption of each gate, respectively, at nominal conditions (nom) and under WF fluctuations from 1% to 5%. As expected, there is an increase in the sensitivity of logic cells to the process variation (σ/μ) when higher levels of WF fluctuations were explored for both metrics. Logic cells with a larger number of inputs are less robust to WF fluctuations. For example, the NAND4 is at least 9.2% and 20.6% more sensitive than NAND2 to the delay and power variabilities, respectively.

The NOR cells are more impacted by process variations than NAND cells. Concerning the complex cells, the AOI21 and OAI211 cells are more robust to delay variability than the OAI21 and AOI211 cells. These cells have similar behavior to the power variability, except when the AOI21 cell suffers 1–3% of deviation, and the OAI211 cell has 5% of variation from nominal values. In general, FinFET logic cells are more sensitive to delay variability for deviations up to 4%, but an opposite behavior can be verified for variations from 5%, i.e., the logic cells become more susceptible to power variability. Moreover, variations of 5% almost triple the sensitivity of logic cells to power variability when compared with 4% of deviation. The mean (μ) of MC simulations at standard version is considered later to estimate the penalties imposed by circuit-level mitigation techniques.

The impact of delay variability can also be evaluated using the higher σ/μ relation among the timing arcs of each logic cell. As shown in Table 6.3, except for the inverter, the sensibilities of all logic cells become more prominent. Moreover, the mean of MC simulations is smaller. This happens because the higher σ/μ relation among the timing arcs normally not corresponds to the worst-case delay. However, most statements previously presented using the σ/μ relation of worst-case delay may still be considered. The main difference is that the AOI21 cell is less sensitive than OAI21 only with 1% of deviation.

6.3   Transistor Reordering

Transistor reordering is a simple technique based on rearranged transistor networks keeping the same logic function. Different transistor combinations change the electrical and physical characteristics of logic cells, and consequently, it also modifies the susceptibility to process variations and radiation-induced soft errors. As an example, Fig. 6.6 shows the schematic and layout of the two possible FinFET implementations of AOI21 cell with three fins. The reordering of transistor

Table 6.1 Propagation delay at nominal conditions and under WF fluctuations adopting as metric the σ/μ relation of worst-case delay

Gates	nom (ps)	1%		2%		3%		4%		5%	
		μ (ps)	σ/μ (%)	μ (ps)	σ/μ (%)	μ (ps)	σ/μ (%)	μ (ps)	σ/μ (%)	μ (ps)	σ/μ (%)
INV	6.3	6.3	4.42	6.4	10.46	6.6	18.01	6.9	25.58	7.2	34.02
NAND2	9.6	9.6	4.02	9.8	9.25	10.0	15.93	10.4	22.51	10.7	29.10
NAND3	14.2	14.2	4.43	14.5	11.12	14.9	18.26	15.4	24.76	16.0	31.35
NAND4	19.2	19.3	4.86	19.7	12.45	20.4	19.44	21.1	25.65	21.8	32.05
NOR2	12.6	12.8	6.79	13.2	16.21	13.7	23.68	14.2	30.91	14.8	39.90
NOR3	19.9	20.3	9.32	21.1	18.06	21.8	24.70	22.5	31.64	23.4	40.64
NOR4	28.1	28.9	10.74	29.9	18.34	30.8	24.64	31.7	31.54	32.9	40.60
AOI21	14.1	14.2	6.87	14.7	15.83	15.2	23.10	15.7	30.35	16.4	39.53
AOI211	21.9	22.4	9.10	23.2	17.48	23.9	24.12	24.7	31.19	25.6	40.45
OAI21	14.2	14.3	6.99	14.8	15.97	15.3	23.25	15.8	30.55	16.5	39.80
OAI211	15.7	15.8	7.33	16.4	16.08	16.9	23.23	17.4	30.58	18.1	40.02

Table 6.2 Power consumption at nominal conditions and under WF fluctuations

Gates	nom (nW)	1%		2%		3%		4%		5%	
		μ (nW)	σ/μ (%)	μ (nW)	σ/μ (%)	μ (nW)	σ/μ (%)	μ (nW)	σ/μ (%)	μ (nW)	σ/μ (%)
INV	427	431	3.39	434	7.07	442	11.92	455	20.84	485	51.83
NAND2	540	534	2.92	539	6.13	547	10.22	562	18.26	596	49.19
NAND3	591	607	3.56	613	7.42	624	12.18	643	21.28	689	56.05
NAND4	663	668	4.09	676	8.45	689	13.79	714	23.95	769	61.97
NOR2	532	543	3.38	548	6.77	556	10.95	571	18.98	606	49.20
NOR3	626	628	3.98.	634	8.21	645	13.20	666	22.05	711	53.28
NOR4	691	698	4.64	707	9.52	722	15.24	747	24.90	803	57.51
AOI21	615	642	3.34	648	6.58	658	10.79	676	18.90	718	49.89
AOI211	649	651	3.93	658	8.10	670	13.22	692	22.77	743	58.03
OAI21	575	576	3.21	580	6.37	589	10.52	606	19.32	646	55.40
OAI211	605	606	3.35	611	6.92	621	11.50	640	21.46	689	62.20

Table 6.3 Propagation delay at nominal conditions and under WF fluctuations adopting as metric the higher σ/μ relation

Gates	nom (ps)	1%		2%		3%		4%		5%	
		μ (ps)	σ/μ (%)	μ (ps)	σ/μ (%)	μ (ps)	σ/μ (%)	μ (ps)	σ/μ (%)	μ (ps)	σ/μ (%)
INV	6.3	6.3	4.42	6.4	10.46	6.6	18.01	6.9	25.58	7.2	34.02
NAND2	7.1	7.1	4.61	7.2	10.83	7.4	18.15	7.7	25.50	8.0	33.98
NAND3	12.4	12.5	4.88	12.8	12.23	13.2	20.06	13.7	27.05	14.3	34.49
NAND4	15.9	16.0	5.58	16.5	14.43	17.2	22.42	17.8	29.33	18.6	36.38
NOR2	11.8	11.9	6.79	12.4	17.14	12.9	24.94	13.4	32.42	14.0	41.70
NOR3	17.5	18.0	10.36	18.8	19.96	19.5	27.11	20.2	34.53	21.1	44.16
NOR4	23.5	24.4	12.51	25.4	21.12	26.2	28.11	26.2	35.71	28.3	45.70
AOI21	11.8	11.9	7.30	12.4	17.17	12.9	24.97	13.4	32.45	14.0	41.75
AOI211	17.6	18.0	10.38	18.8	19.98	19.5	27.13	20.2	34.55	21.1	44.17
OAI21	13.4	13.5	7.35	14.0	16.77	14.5.	24.35	15.0	31.89	15.6	41.45
OAI211	15.2	15.0	7.67	15.5	16.81	16.1	24.22	16.6	31.80	17.3	41.54

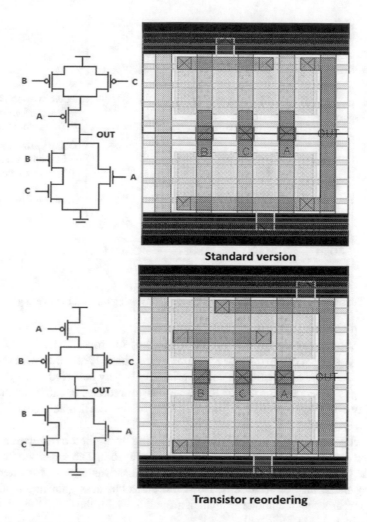

Fig. 6.6 Schematic and layout of the AOI21 cell implemented at standard version and using the transistor reordering technique

"A" maintains the same layout width, increasing the amount of metal 1 (M1) to correctly make the new connections. The reordering of transistors "B" and "C" is not necessary because it does not change the electrical behavior of the AOI21 logic cell. In this way, we did the same reordering for other complex cells, such as OAI21, AOI211, and OAI211. Both implementation versions of these logic cells also keep the same layout width.

Figure 6.7 shows the sensitivity of logic cells designed at standard version and using the transistor reordering technique, under 5% of WF deviation from nominal conditions. For the delay variability, the metric used was the σ/μ relation of worst-case delay. The placement of serial transistors as far as possible to the cell output

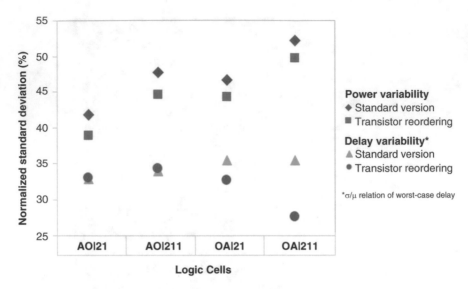

Fig. 6.7 Sensitivity of logic cells to the process variability using transistor reordering

(like the version presented in the right of Fig. 6.6) improves at least 4.9% of the robustness of logic cells to the power variability, reaching 7% for the AOI21 cell. On the other hand, the attenuation of delay variability is only achieved if the transistor reordering is applied on the OAI cells. The transistor reordering enhanced the delay variability robustness around 8% and 22.1% for the OAI21 and OAI211 cells, respectively.

The delay variability can also be measured using the highest σ/μ relation among all timing arcs of each logic cell, as illustrated in Fig. 6.8. The sensitivity tendency is the same. The transistor reordering remains disadvantageous for attenuating the delay variability of AOI cells. However, the robustness gain in adopting the transistor reordering technique is smaller, around 5.6% for the OAI21 logic cell and 10% for the OAI211 logic cell.

The influence of transistor reordering technique on the process variability mitigation was published in [3]. The results are obtained by adopting the HSPICE from Synopsys and the first version of ASAP7 PDK. As all other techniques evaluated in this work used the SPECTRE from Cadence and the last version of ASAP7 PDK, the experiments were re-simulated to ensure a fair comparison. For these reasons, there are some differences between the values presented here and in the paper, but the design flow is the same.

The transistor reordering technique does not add any extra transistor, such that all modifications happen inside the pull-up or pull-down networks. For this reason, the influence of the number of fins has not been evaluated. Moreover, when this technique was evaluated with levels of WF fluctuations varying from 1% to 4%, the improvements in power and delay variability were less than 2%. Thus, as statistically is not advantageous to apply this technique for these variation levels, the results are not presented in this book.

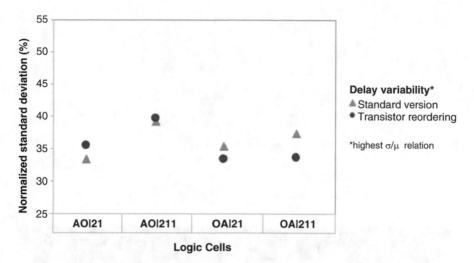

Fig. 6.8 Sensitivity of logic cells to the delay variability using transistor reordering

6.4   Decoupling Cells

The insertion of decoupling cells is a capacitive method that increases the critical charge of the node connected to them. Moreover, the transistors in cross-coupled mode ensure better signal integrity, decreasing the impact caused by process variations. Considering that, for the logic gates investigated, the output node is most vulnerable to soft errors, and the sensitivity of gates to process variability is measured using the output signal, the decoupling cells were connected to the output node. Figure 6.9 shows the schematic and the layout of the AOI21 logic gate using decoupling cells with sizing equal to three fins. This technique demands four extra transistors, increasing the layout width, and consequently, the area of the AOI21 gate around $1.4\times$.

The delay and power variabilities of all logic gates in the standard version and using decoupling cells is shown in Figs. 6.10 and 6.11, respectively, considering 5% of deviation from nominal values and the σ/μ relation of the worst-case delay. According to the results, the adoption of decoupling cells is an effective approach to obtain logic gates more robust to process variations, presenting more significant reductions for power variability.

On average, a design with decoupling cells decreases the delay variability of logic gates around 5.1%, if compared with the standard version. The improvements for the NAND and AOI gates increase as the number of inputs also increases, but the opposite behavior was verified for the NOR and OAI gates, with a drop in the gains. Except for the AOI21 gate, all the others presented a minimal improvement of 4% in the delay variability.

The gains on power variability using decoupling cells are 8.6%, on average. Logic gates with a smaller number of inputs presented higher power variability

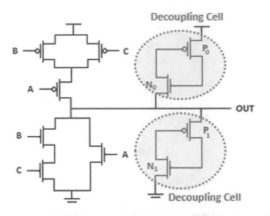

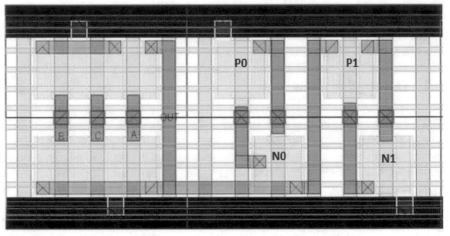

Fig. 6.9 Schematic and layout of AOI21 gate implemented using decoupling cells

mitigation. Among all logic gates evaluated, the NAND2 presented the highest gain (9.4%) in the power variability, while the lowest improvement is equal to 7%, obtained by the NOR4 logic gate.

6.4.1 Variation Levels Impact

The design of each logic gate using decoupling cells was evaluated under different levels of WF fluctuations. As an example, Fig. 6.12 shows the delay variability of NAND2 gate with process variations from 1% to 5% adopting the highest σ/μ relation. The insertion of decoupling cells becomes advantageous for fabrication

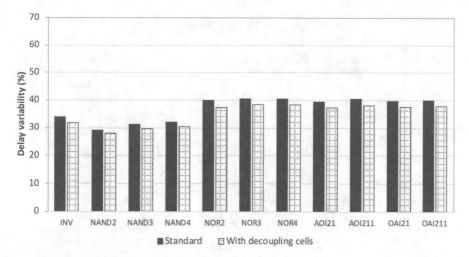

Fig. 6.10 Delay variability using decoupling cells

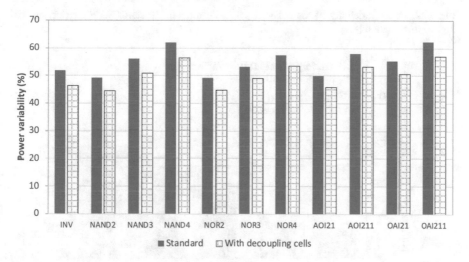

Fig. 6.11 Power variability using decoupling cells

processes with WF fluctuations above 2%. The same behavior was observed for all logic gates studied in this work.

The gains for the NAND2 gate increase as the percentage of variation also increases, reaching up to 5.6% with 5% of WFF, as can be seen in the values highlighted above the circles. This tendency was also verified for the INV and NAND3 cells. However, the opposite happens for all the other gates, such that the variations of 3% instead of 5% presented the most robust results. According to Fig. 6.13, it is possible to analyze the delay variability of the NAND2 gate considering the σ/μ relation of worst-case delay. Although the trend for all logic

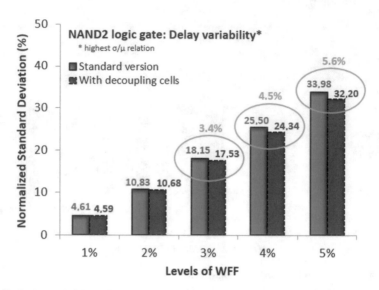

Fig. 6.12 Delay variability of the NAND2 gate with different levels of WFF considering the highest σ/μ relation [4]

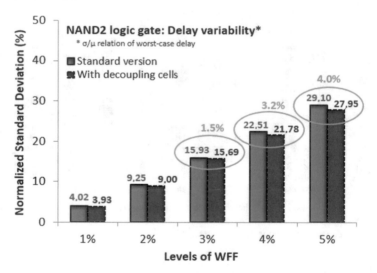

Fig. 6.13 Delay variability of the NAND2 gate with different levels of WFF considering the σ/μ relation of worst-case delay

gates remains the same, the gains in adopting the decoupling cells technique become smaller.

On the other hand, a design with decoupling cells is only advantageous for power variability if the variation from nominal conditions is above 4%, as highlighted in Fig. 6.14. If the NAND2 logic gate suffers a deviation of 5%, the use of decoupling

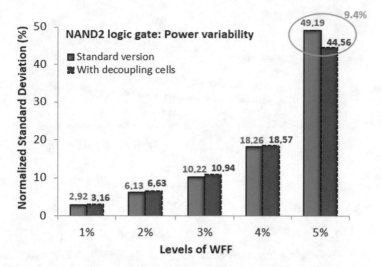

Fig. 6.14 Power variability of the NAND2 gate with different levels of WFF [4]

cells brings an improvement of around 9.4%. Except for the inverter, all the other logic gates presented the same tendency to process variability mitigation. Depending on the levels of WFF, a design with decoupling cells introduced gains in the delay and power variabilities up to 10.3% and 10.7%, respectively.

6.4.2   Sizing Influence

The total capacitance in the output of each gate depends on the transistor sizing of decoupling cells. As FinFET devices have a discrete sizing, it is possible to place multiple fins in parallel to obtain wider FinFETs. The gains of a NAND2 logic gate with decoupling cells exploring a different number of fins can be seen in Fig. 6.15. Process variability robustness increases even more, when larger decoupling cells are used. For example, considering the process variations of 5% WFF, the adoption of decoupling cells with five fins reduces the sensitivity of power and delay variabilities in 3.7% and 2.4% when compared with a layout with three fins, respectively.

On the other hand, the adoption of decoupling cells with 3% of WF variations from nominal values generates a worsening in the sensitivity of power variability. This behavior intensifies when larger decoupling cells are used in the design. The behavioral trend for the other logic gates remains the same.

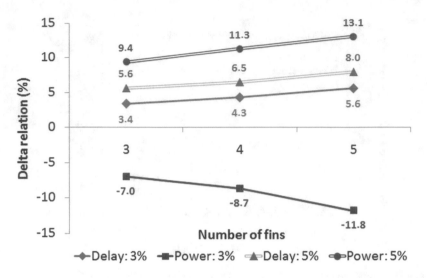

Fig. 6.15 Improvements in connecting decoupling cells with different number of fins in the output of the NAND2 logic gate [4]

6.4.3   Effectiveness of Transistor Reordering

The effectiveness of applying the transistor reordering technique with the insertion of decoupling cells is presented in Figs. 6.16 and 6.17 for the delay and power variabilities, respectively. In this case, the transistor reordering (reor) presents the lowest values in most cases observing the σ/μ relation. The standard topology (std) has a little advantage for the OAI gates under 3% of WFF for delay analysis and the AOI211 complex gate exposed to 5% of WFF. However, in general, the results differ less than 2%, signalizing a not statistically significant difference in the deviation. Thus, the transistor arrangement influence does not show a direct relation to increase the process variability robustness when the decoupling cells are used in the design.

6.5   Schmitt Trigger

The use of Schmitt Triggers is an effective method for increasing the I_{ON}-to-I_{OFF} current ratio and, consequently, for minimizing the output degradation. Moreover, the design with Schmitt Trigger increases the capacitance of the output node of logic cells. These features help to mitigate the process variability effects and the soft error susceptibility. Figure 6.18 illustrates the schematic and layout of the AOI21 cell with a Schmitt Trigger of three fins connected in the output. The six extra transistors imposed by this technique alter the layout width, increasing the area of the AOI21

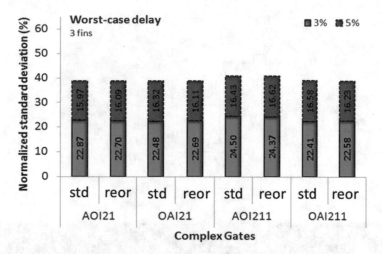

Fig. 6.16 Delay variability exploring decoupling cells with transistor reordering [4]

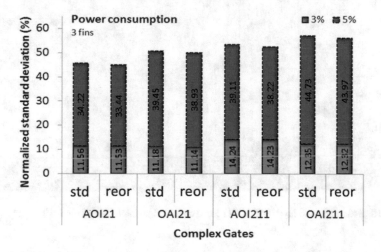

Fig. 6.17 Power variability exploring decoupling cells with transistor reordering [4]

cell around $1.4\times$. Among all methods evaluated, the layout of the Schmitt Trigger is unique to use M2 rails for connecting the source terminals of P_F and N_F transistors.

The sensitivity of logic cells to process variations decreases considerably, connecting the Schmitt Trigger in the output node. The attenuation in the delay and power variabilities using this approach can be seen in Figs. 6.19 and 6.20, respectively, with a fabrication process deviation of 5%. On average, the delay variability of logic cells has an improvement of around 26.6%. The NAND4 is the most benefited cell with the Schmitt Trigger technique, reaching 29.6% of delay variability mitigation.

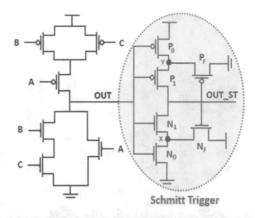

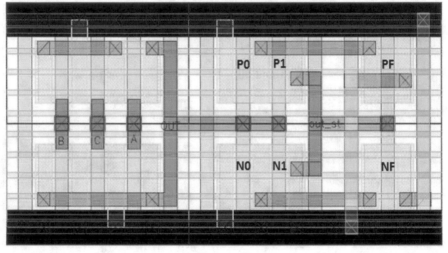

Fig. 6.18 Schematic and layout of AOI21 logic cell implemented using Schmitt Trigger

The design with a Schmitt Trigger provides the power variability mitigation of 28.3%, on average. More improvements can be found for logic cells with a larger number of inputs. However, the minimum gain (19.3%) in power variability adopting this approach is already a considerable value.

6.5.1 Variation Levels Impact

According to the level of WF fluctuation that a logic cell is exposed to, the advantages in adopting the Schmitt Trigger change. To exemplify this statement, Fig. 6.21 shows the impact of the process variations on the delay variability

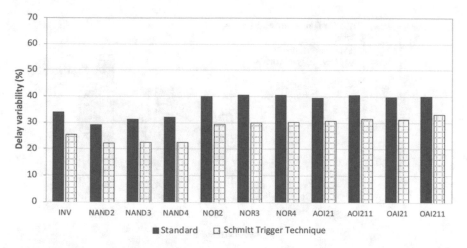

Fig. 6.19 Delay variability mitigation using Schmitt Trigger

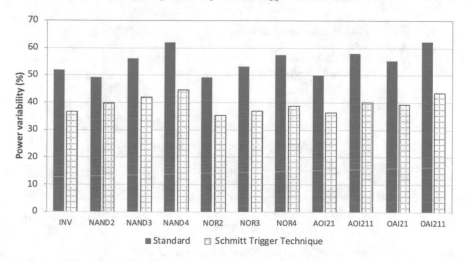

Fig. 6.20 Power variability mitigation using Schmitt Trigger

of AOI21 cell. Lower levels of WF fluctuations improve even more the delay variability. However, logic cells under 1% and 2% of deviation do not follow this trend. This behavior is verified for all logic cells investigated in this work.

Figure 6.22 illustrates the tendency of power variability mitigation when the AOI21 cell connected to a Schmitt Trigger was analyzed under different levels of WF fluctuation. It is possible to verify an opposite behavior of delay variability, such that more improvements in the power variability are achieved with higher levels of deviation. This conclusion for the AOI21 cell can also be applied to all logic cells. On average, the Schmitt Trigger technique provides power variability mitigations of 18% and 22.3% for 3% and 4% of deviation from nominal conditions.

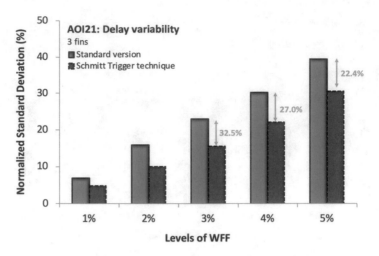

Fig. 6.21 Delay variability of AOI21 cell with different levels of WFF

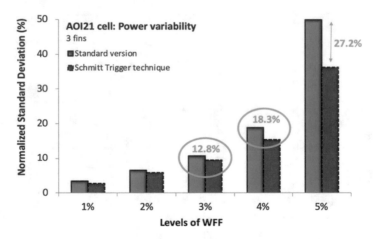

Fig. 6.22 Power variability of AOI21 cell with different levels of WFF

6.5.2 Sizing Influence

The influence in connecting a Schmitt Trigger with a different number of fins in the output of AOI21 cell was verified in Fig. 6.23. Larger Schmitt Triggers decrease the impact on power variability considerably. For example, considering the process variations of 3% and 5%, the use of Schmitt Trigger with five fins improves the sensitivity of power variability in 22.8% and 10.8% if compared with a layout with three fins, respectively. On the other hand, the increase in the number of fins contributes less than 2% for delay variability mitigation, independently of WF fluctuation levels. In this way, investing in larger Schmitt Triggers is only advantageous to attenuate power variability.

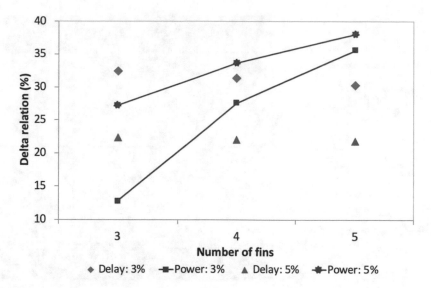

Fig. 6.23 Improvements in connecting a Schmitt Trigger with different number of fins in the output of AOI21 logic cell

6.6   Sleep Transistor

A design with a sleep transistor helps to reduce the leakage currents, transient faults, process variations, and NBTI effects. For all logic gates investigated, a sleep transistor was inserted between the pull-down network and the ground rail. As an example, Fig. 6.24 shows the schematic and layout of AOI21 cell using a sleep transistor with transistor sizing equal to three fins. The extra transistor modifies the layout width, increasing the area of the AOI21 cell around $0.3\times$.

Figures 6.25 and 6.26 show the sensibility of logic gates to the delay and power variabilities, respectively, when a sleep transistor with three fins is added in the design with 5% of WF fluctuation. The results confirm the efficiency of this technique to mitigate the effects of process variation because the σ/μ relation of logic cells is reduced in most of the cases.

The NAND2 cell obtained 17.1% of attenuation in the delay variability. Fewer gains were observed for the NAND cell with three (15.8%) and four (13.7%) inputs. The most significant improvement in the delay variability can be observed in the AOI21 and OAI cells, where the mitigation is 36.4%, on average, for 5% of WF deviation. Nevertheless, the sleep transistor is not advantageous to control the delay variability for the inverter, AOI211, and NOR cells. This worsening is related to how the transistors in the pull-down network are arranged along with the sleep transistor.

On the other hand, a design with a sleep transistor is favorable to mitigate the power variability of all logic gates. Except for the inverter, the gains adopting this

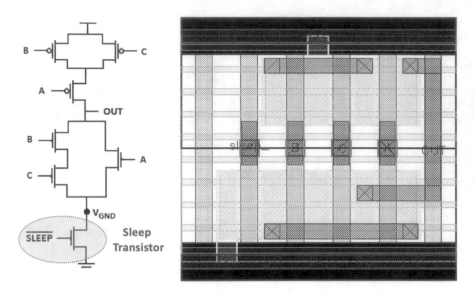

Fig. 6.24 Schematic and layout of AOI21 cell implemented using sleep transistor

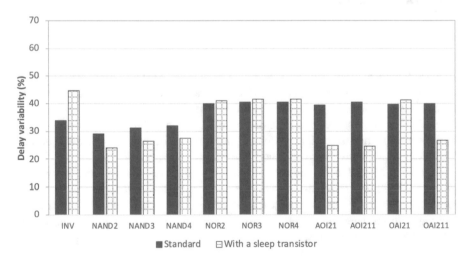

Fig. 6.25 Dclay variability using a sleep transistor

technique vary from 7.9% to 12.4%, considering 5% of deviation. Moreover, the basic (NAND/NOR) and complex (AOI/OAI) cells with a larger number of inputs present more benefits regarding power variability reduction.

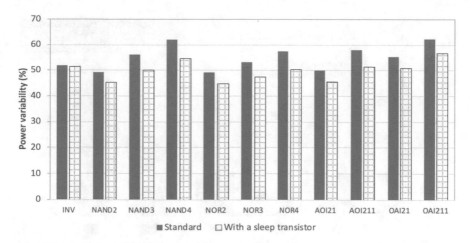

Fig. 6.26 Power variability using a sleep transistor

6.6.1 Variation Levels Impact

The design of each logic cell using a sleep transistor was evaluated under different levels of WF fluctuations (1–5%). Higher levels of WF variations intensify the power variability mitigation, as shown in Fig. 6.27, for the AOI21 logic cell. According to the green values, the gains of AOI21 cell with the sleep transistor technique reach up to 8.5% with 5% of WFF. The same behavior was verified for most of the logic gates, except INV, NOR2, and OAI211 cells. On average, the power variability is improved around 6.5%, and 9.2% for 3%, and 5% of deviation from nominal conditions.

The opposite behavior was verified for the delay variability, such that low levels of WF fluctuations provide better mitigation results for the AOI21 cell, as shown in the green values in Fig. 6.28. The sleep transistor technique improves the delay variability around 51.8% for process variation with 1% of deviation from nominal values. This trend is in agreement with those observed for the Schmitt Trigger technique. However, it is important to highlight that for the sleep transistor technique, this tendency is not valid for all the logic cells studied in this work. Moreover, the addition of a sleep transistor in the design is disadvantageous to mitigate the delay variability of the NOR and AOI211 cells.

6.6.2 Sizing Influence

The influence of different sleep transistor sizes on the process variability mitigation is shown in Fig. 6.29, using the AOI21 cell as an example. Independent of the variation levels, larger sleep s transistors contribute less than 2% both for power

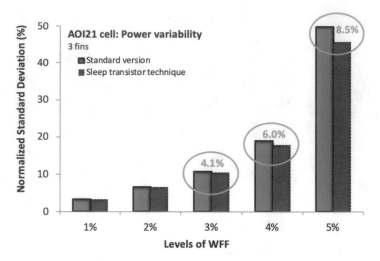

Fig. 6.27 Sensitivity of AOI21 cell to power variability using a sleep transistor with different levels of WFF [5]

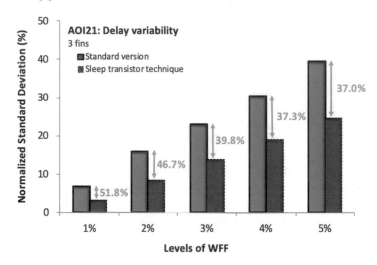

Fig. 6.28 Sensitivity of AOI21 cell to delay variability using a sleep transistor with different levels of WFF [5]

and delay mitigations, if compared with the smaller version (3 fins). This behavior is similar to all the other logic cells adopting different transistor sizing. In this way, the best alternative is to use a smaller sleep transistor, ensuring favorable process variability mitigation, avoiding even more penalties in performance, power consumption, and area.

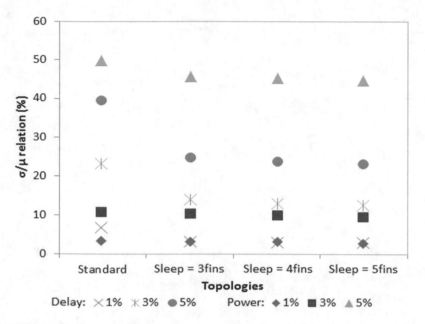

Fig. 6.29 Impact of sleep transistor sizing in the process variability mitigation of AOI21 gate [5]

References

1. Brendler, L.H., Zimpeck, A.L., Meinhardt, C., Reis, R.: Multi-level design influences on robustness evaluation of 7nm FinFET technology. IEEE Trans. Circuits Syst. I: Regular Papers **67**(2), 553–564 (2020)
2. Meinhardt, C.: Variabilidade em FinFETs. Thesis (Doutorado em Ciência da Computação) – Instituto de Informática - UFRGS (2014)
3. Zimpeck, A., Meinhardt, C., Artola, L., Hubert, G., Kastensmidt, F., Reis, R.: Impact of different transistor arrangements on gate variability. Microelectron. Reliab. **88–90**, 111–115 (2018). 29th European Symposium on Reliability of Electron Devices, Failure Physics and Analysis (ESREF 2018)
4. Zimpeck, A., Meinhardt, C., Artola, L., Hubert, G., Kastensmidt, F., Reis, R.: Mitigation of process variability effects using decoupling cells. Microelectron. Reliab. **100–101**, 113,446 (2019). 30th European Symposium on Reliability of Electron Devices, Failure Physics and Analysis
5. Zimpeck, A.L., Meinhardt, C., Artola, L., Hubert, G., Kastensmidt, F.L., Reis, R.: Sleep transistors to improve the process variability and soft error susceptibility. In: 2019 26th IEEE International Conference on Electronics, Circuits and Systems (ICECS), pp. 582–585 (2019)

Chapter 7
Soft Error Mitigation

Soft errors are transient events with a short time interval induced by energetic particles coming from terrestrial and space radiations. Radiation-induced soft errors may cause critical failures in system behavior, leading to financial or human life losses. This chapter explores the behavior of FinFET logic cells at nominal conditions and also under radiation-induced soft errors. The standard version of logic cells was evaluated without any circuit-level mitigation technique. In general, all approaches demonstrate interesting results to control the soft error susceptibility.

7.1 Standard Design

This section evaluates the impact of soft errors in three FinFET logic cells under low LET values, i.e., less than $15 \, \text{MeV} \, \text{cm}^2 \, \text{mg}^{-1}$. These LET values have been targeted because they correspond to secondary particles induced by neutrons at avionic and ground applications. The metric for soft error evaluation is the SET cross section, considering the most sensitive input vector and also the mean of all input vectors.

Figure 7.1 shows the soft error susceptibility of NAND2, NOR2, and AOI21 cells at near-threshold regime (0.3 V) considering the most sensitive input vector. For all LETs investigated, the three logic gates are free of faults at 0.6 V and core voltage. The AOI21 cell is free of faults with a LET of $5 \, \text{MeV} \, \text{cm}^2 \, \text{mg}^{-1}$, and soft errors are only seen in the output of NAND2 and NOR2 cells at 0.3 V. On the other hand, it is possible to observe some faults at 0.4 and 0.5 V when higher LET ($15 \, \text{MeV} \, \text{cm}^2 \, \text{mg}^{-1}$) was investigated. The reduction of the number of faults happens due to the FinFET disruptive nature that increases the minimum charge required to induce a SET pulse. Moreover, this behavior is consistent with the previous work obtained for the same technology on majority voters [1].

The logic cell more susceptible to soft errors is the AOI21 because most of the current sources from the SET database generated visible faults at the output.

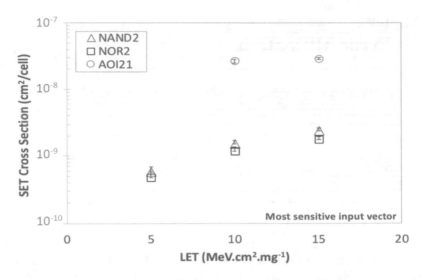

Fig. 7.1 SET cross section of logic cells operating at near-threshold regime considering the most sensitive input vector

However, the LET increase from 10 to $15\,\mathrm{MeV\,cm^2\,mg^{-1}}$ does not significantly impact the cross section. On the other hand, the increase of the cross section of basic cells is almost linear to the increase of LET according to Fig. 7.1. The simulations indicate a higher SET cross section to the NAND2 cell, such that it also agrees with the previous work that uses the ASAP7 model [1]. The NOR2 cell is around 17.4%, 21.7%, and 25.5% more robust than the NAND2 cell to soft error impact for LETs equal to 5, 10, and $15\,\mathrm{MeV\,cm^2\,mg^{-1}}$, respectively.

Figure 7.2 demonstrates the same analysis, but the cross section was calculated using the mean of faults for all input vectors. It is possible to note that the susceptibility of basic cells to soft error is very similar when all input vectors were considered. Moreover, the SET cross section of the AOI21 cell decreases around 65% for both LETs because the other input vectors manifested fewer faults than the most sensitive input vector. The error bars in all SET cross section graphs are defined as one divided by the square root of the number of SETs representing the statistical error induced by Monte Carlo simulations [2].

7.2   Transistor Reordering

A comparison between the SET cross section of AOI21 cell considering the standard version and the transistor reordering technique is shown in Fig. 7.3, for most sensitive input vector and a LET of $10\,\mathrm{MeV\,cm^2\,mg^{-1}}$. For this LET, both topologies show no events on the range from 0.5 V to the nominal voltage (0.7 V).

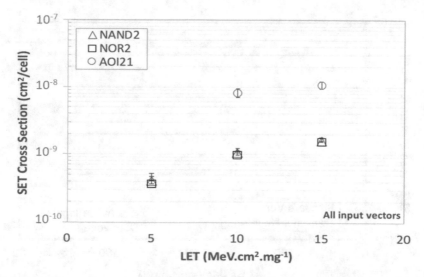

Fig. 7.2 SET cross section of logic cells operating at near-threshold regime considering all input vectors

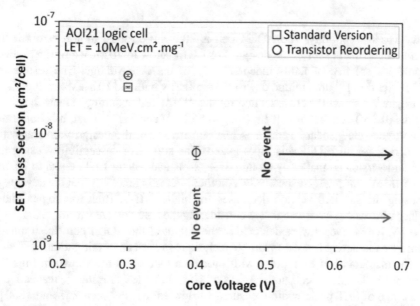

Fig. 7.3 SET cross section of AOI21 cell under a LET of $10\,\mathrm{MeV\,cm^2\,mg^{-1}}$ considering most sensitive input vector [3]

Consequently, the choice of different transistor arrangements in the AOI21 cell for these supply voltages does not influence its susceptibility to soft errors.

At a near-threshold regime (0.3–0.4 V), the transistor reordering putting the serial transistors as far as possible to the output is not favorable to mitigate the

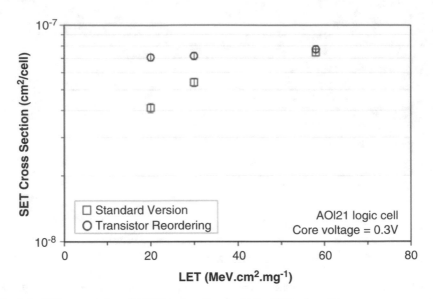

Fig. 7.4 SET cross section of AOI21 logic cell under higher LET values [3]

soft error susceptibility. The AOI21 cell is 20% more robust to soft errors if the standard version is kept in the design. For LETs below to $10 \, \mathrm{MeV \, cm^2 \, mg^{-1}}$, both topologies are free of faults independently of the core voltage. The results for $15 \, \mathrm{MeV \, cm^2 \, mg^{-1}}$ are omitted due to the similarity with the $10 \, \mathrm{MeV \, cm^2 \, mg^{-1}}$.

Figure 7.4 shows the cross section for the AOI21 cell operating at near-threshold regime (0.3 V) as a function of LETs up to $58 \, \mathrm{MeV \, cm^2 \, mg^{-1}}$. First, it is important to note that the standard version is less sensitive than the design with transistor reordering for all LET levels investigated. The soft error susceptibility using the standard version is smaller by around 37%, 24%, and 5% for LETs equal to 20, 30, and $58 \, \mathrm{MeV \, cm^2 \, mg^{-1}}$, respectively. Another interesting factor is that the advantage in using the standard version decreases for higher LETs. Both topologies under higher LET values were also tested considering the core voltage varying from 0.4 V to 0.7 V, but no events were detected in the output of the AOI21 cell. In summary, transistor reordering is not a favorable technique for soft error mitigation.

The histograms of SET pulse width for both topologies are shown in Figs. 7.5 and 7.6 for the AOI21 cell under LETs of 10 and $58 \, \mathrm{MeV \, cm^2 \, mg^{-1}}$, respectively. The range of SET pulse width measured for low LET is between 267 and 1800 ps while for high LET dose is between 120 and 600 ns. As expected, the SET pulse widths have a wide distribution and increase significantly for higher LET levels.

The standard version generates predominantly smaller SET pulse widths than the full distribution of pulses in both LETs investigated. For a LET of $10 \, \mathrm{MeV \, cm^2 \, mg^{-1}}$, most of SET pulse widths (around 90%) are smaller than 1000 ps. The opposite happens with a design using the transistor reordering

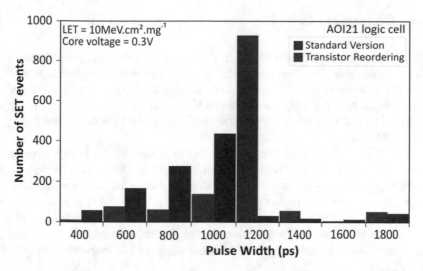

Fig. 7.5 SET pulse width distribution for the AOI21 gate designed in the standard version and using the transistor reordering under a LET of 10 MeV cm² mg⁻¹ [3]

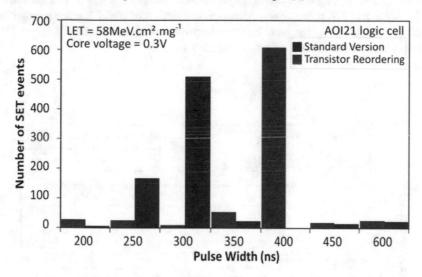

Fig. 7.6 SET pulse width distribution for the AOI21 gate designed in the standard version and using the transistor reordering under a LET of 58 MeV cm² mg⁻¹ [3]

technique, such that 78% of the SET pulse widths are larger than 1000 ps. For a LET of 58 MeV cm² mg⁻¹, a similar tendency can be observed. The transistor reordering results in 92% of the SET pulse widths larger than 350 ps.

7.3   Decoupling Cell

The SE susceptibility of the NAND2 gate was investigated considering the standard version and connecting decoupling cells in the output with the core voltage varying from 0.7 V down to 0.3 V. From nominal voltage until 0.5 V, the NAND2 gate is free of faults for both design possibilities. Few faults were observed in the output at 0.4 V, i.e., less than 6% of current sources from the SET database considering all LETs evaluated.

Nevertheless, the design using decoupling cells presented a little improvement in the SET vulnerability. The comparison between the SET cross section of the NAND2 gate in the standard version and with the decoupling cells connected in the output at near-threshold regime (0.3 V) is presented in Fig. 7.7. The results showed that the SE vulnerability decreases around 24.5%, 23.7%, and 11.4% for LETs equal to 15, 20, and 58 MeV cm^2 mg^{-1} when decoupling cells are adopted in the design. Like the previous technique, the use of decoupling cells is more advantageous for lower LETs.

The SET pulse width distribution to the NAND2 gate in the standard version and using decoupling cells for LETs of 15 and 58 MeV cm^2 mg^{-1} can be shown in Figs. 7.8 and 7.9, respectively. Although the design of FinFET gates connecting decoupling cells in the output is better to improve the SE susceptibility, larger SET pulses were verified mainly under higher LET levels.

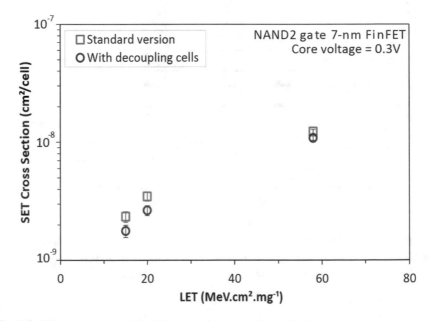

Fig. 7.7 SET cross section of NAND2 gate using decoupling cells [3]

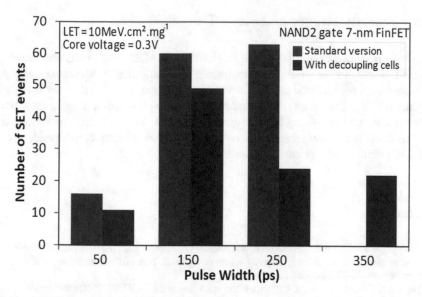

Fig. 7.8 SET pulse width distribution for the NAND2 gate designed with and without decoupling cells under a LET of $10\,\text{MeV}\,\text{cm}^2\,\text{mg}^{-1}$ [3]

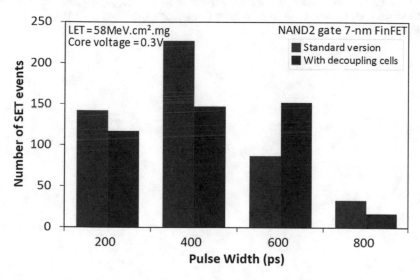

Fig. 7.9 SET pulse width distribution for the NAND2 gate designed with and without decoupling cells under a LET of $58\,\text{MeV}\,\text{cm}^2\,\text{mg}^{-1}$ [3]

7.4 Schmitt Trigger and Sleep Transistor

The soft error susceptibility of NAND2, NOR2, and AOI21 logic cells was analyzed under low and higher LETs, varying from 5 to 58 cm^2 mg^{-1}. Moreover, all input vectors were tested as well as the core voltage was modified until the near-threshold regime (0.3 V). The results show that a design using a Schmitt Trigger or a sleep transistor is very promising for soft error mitigation. The three logic cells become free of faults (no events seen at gate output) using this technique, independently of the LET, input vector, or core voltage employed in the design.

References

1. de Aguiar, Y., Artola, L., Hubert, G., Meinhardt, C., Kastensmidt, F., Reis, R.: Evaluation of radiation-induced soft error in majority voters designed in 7nm FinFET technology. Microelectron. Reliab. **76–77**, 660–664 (2017)
2. Hubert, G., Artola, L.: Single-event transient modeling in a 65-nm bulk CMOS technology based on multi-physical approach and electrical simulations. IEEE Trans. Nuclear Sci. **60**(6), 4421–4429 (2013)
3. Zimpeck, A.L., Meinhardt, C., Artola, L., Hubert, G., kastensmidt, F.L., Reis, R.A.L.: Circuit-level hardening techniques to mitigate soft errors in FinFET logic gates. In: European Conference on Radiation and its Effects on Components and Systems (RADECS), pp. 1–4 (2019)

Chapter 8
General Trade-Offs

The use of circuit-level techniques brought several benefits regarding the mitigation of process variability and soft errors. However, some approaches add extra transistors in the design that consequently increase the area, power consumption, and performance of logic cells compared with the standard version. Moreover, even that the transistor reordering keeps the same number of transistors in the design, this approach modifies the electrical behavior, introducing some penalties in the metrics. The results presented in this chapter seek: (1) to prove the efficiency of the four circuit-level mitigation approaches investigated in this book considering different test scenarios, (2) to indicate the pros and cons in adopting each one of them, and (3) to provide an overall comparison to allow the designers to choose the best technique depending on the target application.

8.1 Technique Drawbacks

8.1.1 Area

All transistors of the logic cells were designed using the same sizing, but the extra transistors imposed by some approaches were evaluated using different sizing to verify the mitigation capability. The number of extra transistors imposed by each technique and the total area of the logic cells using them is shown in Table 8.1.

In terms of area occupied, the transistor reordering is the best technique because it has no area penalties. The addition of decoupling cells or Schmitt Triggers in the cell output introduces a different number of extra transistors, but the area penalty for both is the same. This happens because gate terminal signal is different for the pair of transistors in the cross-coupled mode. Since the ASAP7 PDK does not allow the gate layer break, more area is demanded in the layout of decoupling cells to adjust

© The Author(s), under exclusive license to Springer Nature Switzerland AG 2021
A. Zimpeck et al., *Mitigating Process Variability and Soft Errors at Circuit-Level for FinFETs*, https://doi.org/10.1007/978-3-030-68368-9_8

Table 8.1 Area penalties in adopting each circuit-level mitigation techniques

Techniques	Extra transistors	Szing	Area of logic cells (nm^2)			
			1 input	2 inputs	3 inputs	4 inputs
Standard version	–	3 fins	50.9	67.8	84.8	101.7
Transistor reordering	0	3 fins	–	–	84.8	101.7
Decoupling cells	4	3 fins	169.6	186.5	203.5	220.4
		4 fins	198.7	218.6	238.5	258.3
		5 fins	227.9	250.7	273.5	296.2
Schmitt trigger	6	3 fins	169.6	186.5	203.5	220.4
		4 fins	198.7	218.6	238.5	258.3
		5 fins	227.9	250.7	273.5	296.2
Sleep transistor	1	3 fins	67.8	84.8	101.7[a]	118.7[a]
		4 fins	79.5	99.4	119.2[a]	139.1[a]
		5 fins	91.2	113.9	136.7[a]	159.5[a]

[a]These values change for the OAI logic cells

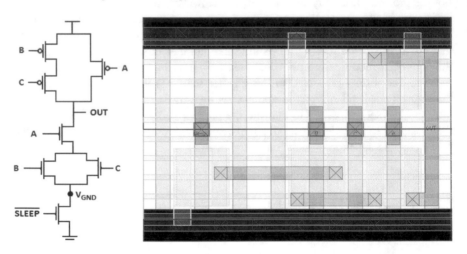

Fig. 8.1 Diffusion break to connect the sleep transistor to the standard design of OAI21 cell

each input (see Fig. 6.9). The increase of area can vary between 2.2× and 4.5×, depending on the number of inputs and the sizing of extra transistors.

The insertion of a sleep transistor generates a small area increase of logic cells that vary from 1.2× to 1.8× in most cases. However, there is an exception to the OAI21 and OAI211 logic cells. As shown in Fig. 8.1 for the OAI21 cell, a diffusion break was required to connect the standard design and the sleep transistor correctly. For each diffusion break, the ASAP7 PDK demands two dummy gates at each end. This constraint increases the estimated area in Table 8.1 around 33 and 28% for OAI21 and OAI211 cells designed with sleep transistors, respectively.

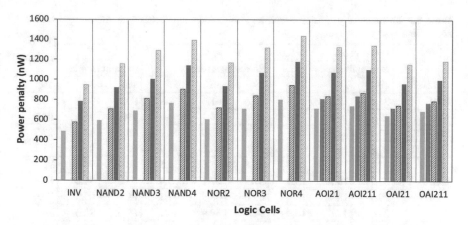

Fig. 8.2 Power penalty of using circuit-level mitigation approaches

8.1.2 Power Consumption

The four circuit-level mitigation techniques explored in this work also introduced penalties in the power consumption of the logic cells. As expected, approaches that add extra transistors bring a higher impact over the power consumption, as shown in Fig. 8.2, considering 5% of deviation and the extra transistor sizing equal to three fins. In this way, a design with Schmitt Trigger connected to the cell output suffers the higher average impact on power (84.6%). The insertion of decoupling cells has four additional transistors. The advantage is that transistors in the cross-coupled mode consume less power. For this reason, the average impact on power due to the addition of decoupling cells is around 17.6%. Although the sleep transistor presented the fewer area penalty, this device connected to the pull-down network introduces a considerable power overhead of 50.4%, on average. Finally, the impact of the transistor reordering approach on power consumption is around 11.1

The power penalties increase even more for low levels of WF fluctuations, as shown in Table 8.2. This behavior can be seen for all mitigation techniques, but the growth is not significant for the designs using a sleep transistor. On average, a design with decoupling cells under 1% of deviation from nominal conditions suffers 12.3% more penalties in the power consumption if compared with 5% of deviation. The use of Schmitt Trigger remains the approach that most impacts the power of logic cells, independently of the level of WF fluctuations.

The extra devices imposed by decoupling cells, sleep transistor, and Schmitt Trigger techniques were evaluated under a different number of fins. As expected, larger devices increase the power penalties, as demonstrates Table 8.3 considering 5% of deviation. Decoupling cells with four and five fins generate an increase in the

Table 8.2 Power penalty of using circuit-level mitigation approaches with different levels of WF fluctuations

| Gates | Power penalties (%)–3 fins | | | | | | | | |
| | Decoupling cells | | | Sleep transistor | | | Schmitt trigger | | |
	1%	3%	5%	1%	3%	5%	1%	3%	5%
INV	31.6	19.7	18.6	60.8	61.3	62.1	99.1	94.1	95.3
NAND2	19.7	19.4	18.8	55.8	54.7	54.7	99.1	94.3	94.0
NAND3	43.7	18.9	18.1	53.2	52.9	45.7	95.1	92.1	87.2
NAND4	29.8	18.4	17.8	50.6	50.2	48.4	90.6	87.2	81.1
NOR2	31.3	19.6	19.0	55.8	55.6	54.5	98.9	96.8	92.7
NOR3	30.1	18.9	18.3	51.9	52.1	50.4	93.8	91.0	85.4
NOR4	29.7	18.4	18.1	49.6	48.9	47.1	89.4	85.5	79.0
AOI21	29.3	17.6	16.9	51.4	50.9	49.7	89.6	88.6	84.8
AOI211	29.2	17.9	17.2	50.7	50.7	48.3	88.0	86.6	80.9
OAI21	27.8	16.3	15.6	50.3	49.9	48.9	83.5	82.7	78.6
OAI211	26.6	15.3	14.8	46.9	46.5	45.1	78.2	77.1	72.1
Average	29.9	18.2	17.6	52.5	52.2	50.4	91.4	88.7	84.6

Table 8.3 Power penalty of using circuit-level mitigation approaches with different number of fins and 5% of deviation

| Techniques | Power penalties (%)–4 and 5 fins | | | | | |
| Sizing | Decoupling cells | | Sleep transistor | | Schmitt trigger | |
	4 fins	5 fins	4 fins	5 fins	4 fins	5 fins
INV	24.1	30.5	62.3	63.7	114.6	136.7
NAND2	24.2	30.5	54.7	54.7	114.1	136.6
NAND3	23.7	30.0	51.8	52.5	106.5	127.9
NAND4	23.0	29.3	48.4	48.8	99.7	120.4
NOR2	24.4	31.0	54.6	55.9	113.5	136.3
NOR3	23.8	30.1	50.4	51.6	105.8	127.7
NOR4	23.4	29.6	47.1	48.2	98.8	120.3
AOI21	21.9	27.9	49.7	51.0	103.8	124.4
AOI211	20.3	25.7	48.7	49.1	100.0	120.9
OAI21	22.3	28.4	48.8	50.0	96.0	114.9
OAI211	19.2	24.4	45.6	46.2	88.4	106.1
Average	22.7	28.9	51.1	52.0	103.7	124.7

power consumption around 5% and 11%, respectively, compared with the three fins version. This percentage of penalty does not rise to lower levels of variation (1–4%).

For the sleep transistor technique, independently of the transistor sizing and the levels of WF fluctuations, the impact on power consumption is almost the same, around 50–53%. The critical case is related to the addition of a larger Schmitt Trigger. On average, each fin added to the design generates a worsening of 20% in

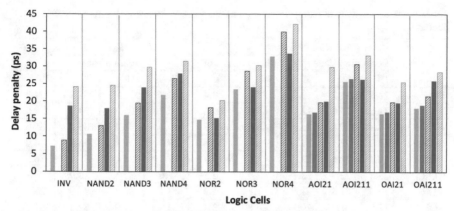

Fig. 8.3 Delay penalty of using circuit-level mitigation approaches

the power consumption of logic cells. More disadvantages can be seen when lower levels of deviation (1–4%) were investigated.

8.1.3 Performance

The extra transistors of the circuit-level mitigation approaches also introduced a performance drop in the logic cells. Figure 8.3 illustrates the delay penalty of all logic cells under 5% of deviation using each technique. The connection of Schmitt Triggers with three fins in the cell output generates an increase of 74.6% in the delay metric, on average. The impact on the delay halved (36.3%) when sleep transistors were adopted instead of the Schmitt Trigger technique observing the average. The delay overhead introduced by using decoupling cells is around 21.2%, so the delay is much less impacted than the power metric. Finally, the transistor reordering technique modifies the performance of complex cells around 3.5%.

Except for the decoupling cells technique, the penalties on delay decrease for low WF fluctuation levels, on average, as shown in Table 8.4. For the AOI211 and NOR cells, the adoption of a sleep transistor introduced a little impact on delay. On the other hand, the insertion of decoupling cells can be more attractive for INV, AOI21, OAI21, and NAND cells to reduce the impact on delay. Independent of the levels of WF fluctuations, the Schmitt Trigger remains the most impacted regarding the performance. Larger sleep transistors decrease the impact on delay, as shown in Table 8.5, considering 5% of deviation. However, if decoupling cells or Schmitt Triggers are connected to the output, the influence on the delay grows as the sizing increases.

Table 8.4 Delay penalty of using circuit-level mitigation approaches with different levels of WF fluctuations

| | Delay penalties (%)–3 fins | | | | | | | | |
| Gates | Decoupling cells | | | Sleep transistor | | | Schmitt trigger | | |
	1%	3%	5%	1%	3%	5%	1%	3%	5%
INV	24.6	24.2	22.2	125.4	142.4	159.7	212.7	233.3	234.7
NAND2	21.6	23.0	22.4	66.0	69.0	68.2	107.2	124.0	129.0
NAND3	37.3	22.8	21.3	48.6	49.7	49.4	78.2	85.2	85.6
NAND4	21.8	22.1	21.6	32.1	32.8	28.0	39.9	43.6	44.0
NOR2	24.2	24.1	23.0	1.6	2.2	2.7	23.4	32.1	36.5
NOR3	23.6	22.9	22.2	2.0	2.3	2.6	27.1	28.9	29.1
NOR4	20.4	21.8	21.3	1.7	1.9	2.4	30.4	29.9	28.0
AOI21	22.5	21.7	20.1	26.8	23.7	22.0	78.9	82.2	81.7
AOI211	25.4	20.9	19.9	1.8	2.5	2.7	28.1	29.7	29.3
OAI21	22.4	21.6	20.0	23.1	20.3	18.8	47.6	52.9	54.5
OAI211	20.3	19.5	18.8	45.6	43.8	43.1	51.9	55.6	56.9
Average	24.0	22.2	21.2	v34.0	35.5	36.3	65.9	72.5	73.6

Table 8.5 Delay penalty of using circuit-level mitigation approaches with different number of fins and 5% of deviation

| Delay penalties (%)–4 and 5 fins | | | | | | |
| Techniques | Decoupling cells | | Sleep transistor | | Schmitt trigger | |
Sizing	4 fins	5 fins	4 fins	5 fins	4 fins	5 fins
INV	29.2	37.5	138.9	140.3	243.9	252.9
NAND2	29.0	36.4	58.9	53.3	137.9	146.9
NAND3	28.1	35.6	41.9	37.5	94.4	103.8
NAND4	27.5	35.3	32.6	29.4	54.1	61.9
NOR2	29.1	37.8	2.0	2.0	45.7	55.1
NOR3	28.2	36.3	2.1	2.1	38.0	47.4
NOR4	27.1	35.0	2.1	1.8	36.5	45.6
AOI21	26.2	34.1	15.2	11.0	86.9	93.9
AOI211	25.8	33.2	2.3	2.0	37.5	49.8
OAI21	26.1	33.9	12.1	7.9	70.9	83.0
OAI211	24.3	31.5	35.9	32.0	68.0	72.4
Average	27.3	35.2	31.3	29.0	83.1	92.1

8.2   Overall Comparison

This subsection summarizes all conclusions obtained about the presented circuit-level techniques presented in this book. Also, this subsection points out the best circuit-level mitigation technique depending on the target application. Logic cells can be submitted to high (5%), medium (3%), or low (1%) levels of WF fluctuations,

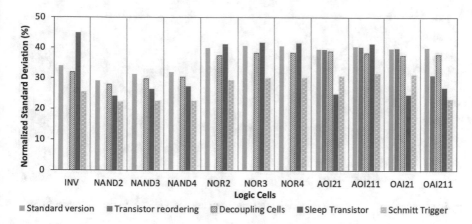

Fig. 8.4 Power variability using circuit-level mitigation approaches

such that these variations can impact the power, propagation delays, or both. The sensitivity of logic cells to the process variations is measured through the normalized standard deviation (σ/μ relation). As previously presented, the deviation on power and propagation delays due to the process variations are denominated in this work as power and delay variability.

Figure 8.4 shows the impact of a fabrication process with 5% of deviation when the standard version or circuit-level mitigation approaches are adopted in the design. The best technique to attenuate the effect on power variability is based on the insertion of Schmitt Triggers. Even for lower levels of deviation, the addition of Schmitt Trigger remains the most advantageous. After that, the most indicated technique to power variability mitigation is the insertion of sleep transistors or decoupling cells. The improvement of both techniques is similar for designs with 5% of deviation. However, for lower levels of WF fluctuation (1–4%), the sleep transistor approach is the second best option. Finally, the technique with fewer gains on power variability attenuation is the transistor reordering.

It is harder for the delay variability mitigation to find a general trend for all logic cells evaluated, as shown in Fig. 8.5. Except for the AOI21 and OAI21 cells, the insertion of a Schmitt Trigger is the best technique to improve the delay variability. The second alternative also varies according to the logic cells. For example, the adoption of a sleep transistor is better than decoupling cells for the NAND and OAI211 cells, but the opposite happens for the INV, NOR, and AOI211 cells. For the AOI21 and OAI21 cells, the first best option is using the sleep transistors, and after, the Schmitt Trigger transistor is more indicated. The transistor reordering technique only brought significant advantages of delay variability mitigation for the OAI211 cell. For lower WF fluctuation levels (1–4%), the same tendency was observed for all logic cells.

The four techniques used to mitigate the effects of process variations and radiation-induced soft errors introduced some penalties regarding area, power

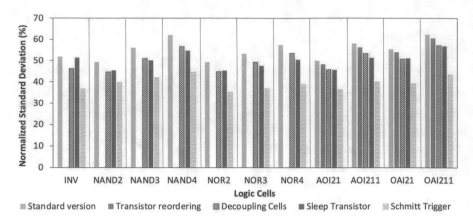

Fig. 8.5 Delay variability using circuit-level mitigation approaches

consumption, and performance. The transistor reordering technique has no area penalties, and the impact on power and performance is small. However, this technique not presented significant improvements in the variability attenuation of complex cells. Moreover, the reordering of transistors becomes the logic cells more sensitive to soft errors.

The connection of Schmitt Trigger in the output of logic cells is the best way to decrease the impact of process variations significantly and also to obtain logic cells free of faults even at the near-threshold regime. However, higher penalties are observed, mainly regarding power consumption and performance. The insertion of decoupling cells has the same area penalty as the Schmitt Trigger technique, but the drawbacks in power consumption and performance are halved. Moreover, a design with decoupling cells decreases the soft error susceptibility of logic gates. The sleep transistor approach introduced low area penalties and logic cells free of faults even at near-threshold regime, but the impact on power consumption and performance is considerable. For most of the techniques and logic cells analyzed, the penalties are higher with lower levels of WF fluctuation.

Decoupling cells with a larger number of fins increase, even more, the robustness of logic cells. However, the penalties increase as the number of fins also increases. But for this technique, the penalties introduced are still agreeable. In the same way, larger Schmitt Triggers in the design increase the power variability mitigation, but the technique drawbacks on power consumption and performance are unacceptable. On the other hand, sleep transistors with a larger number of fins do not significantly minimize the sensitivity of logic cells. For this reason, larger sleep transistors are not advised due to the high penalties involved.

In general, considering the average results of logic cells obtained for all circuit-level mitigation approaches as well as an overall evaluation of all topics and test scenarios presented in this work, it is possible to conclude that:

- Schmitt Trigger is the best technique if the focus of designers is only improving the impact of variability, without any area, performance or power requirements;
- Sleep transistor is the best option if the focus of designers is in increasing the process variability robustness, but they have some area restrictions;
- Decoupling cells are indicated if the focus of designers is in improving the impact of variability, but they have some power or performance requirements;
- Sleep transistor or Schmitt Trigger are more appropriate if the focus of designers is only to decrease the soft error susceptibility;
- Sleep transistor is more indicated if the focus of designers attenuates the impact of soft errors, but they have some area constraints;
- Decoupling cells are the best alternative if the focus of designers is in improving the process variability effects and also become a circuit more robust to transient faults, with acceptable penalties on area, performance, and power consumption.

As previously mentioned in the introduction, few works are exploring circuit-level approaches to mitigate the effects of process variability and soft error in FinFET technologies. Currently, there are four works available in the literature directly related to the subject of this work.

In [4], the traditional inverters of FinFET full adders were replaced by Schmitt Triggers at the layout level, and the process variability sensitivity of these circuits was verified. For most cases evaluated, the adoption of Schmitt Triggers improves the power and delay variability, but with significant overhead, mainly on the area. The results obtained in [4] are in agreement with those presented in this book. However, all overheads are more prominent due to the insertion of more than one Schmitt Trigger in the full adder design.

In [2], different complex cells were designed at the layout level using the multi-level design (only with NAND gates). The process variability and soft error sensitivity were analyzed considering both topologies. Despite the area impact, the multi-level design mitigates at least 50% the delay variability than the version of complex gates. Moreover, the multi-level version improves over 45%, on average, the fault coverage evaluation from SET effects. The improvements obtained in [2] are similar to the Schmitt Trigger technique applied in this work, but the penalties on area, delay, and power are higher. The comparison among the soft error results is not fare, because in [2], the fault injection happens though the double exponential current using SPICE simulations.

In [3] and [1], circuit-level techniques based on the strengthening and redundancy were applied in FinFET circuits to enhance the soft error susceptibility, respectively. Although both approaches demonstrated auspicious outcomes regarding soft error robustness, the experiments consider the estimation using the double exponential and are not considering the layout features. In this way, the comparison with the experimental setup adopted in this book also is not suitable.

References

1. Alghareb, F.S., Ashraf, R.A., Alzahrani, A., DeMara, R.F.: Energy and delay tradeoffs of soft-error masking for 16-nm FinFET logic paths: survey and impact of process variation in the near-threshold region. IEEE Trans. Circuits Syst. II: Express Briefs **64**(6), 695–699 (2017)
2. Brendler, L.H., Zimpeck, A.L., Meinhardt, C., Reis, R.: Exploring multi-level design to mitigate variability and radiation effects on 7nm FinFET logic cells. In: 2018 25th IEEE International Conference on Electronics, Circuits and Systems (ICECS), pp. 581–584 (2018)
3. Calomarde, A., Amat, E., Moll, F., Vigara, J., Rubio, A.: Set and noise fault tolerant circuit design techniques: Application to 7nm FinFET. Microelectron. Reliab. **54**(4), 738–745 (2014)
4. Moraes, L., Zimpeck, A., Meinhardt, C., Reis, R.: Evaluation of variability using schmitt trigger on full adders layout. Microelectron. Reliab. **88–90**, 116–121 (2018). 29th European Symposium on Reliability of Electron Devices, Failure Physics and Analysis (ESREF 2018)

Chapter 9
Final Remarks

FinFET devices were widely adopted by the semiconductor industry for technology nodes sub-22nm, raising essential topics related to the reliability of electronic systems. The small geometric patterns imposed by advanced technologies intensify the process variations, and the higher density allows that a single energetic particle affects multiple adjacent nodes.

The main consequences of these challenges are the parametric yield loss, and the critical failures on system behavior, leading to financial or human life losses. The impact of process variability continues to increase at each new technology node, becoming harder to keep the technology scaling down using FinFET devices.

From a design standpoint, process variations and radiation-induced soft errors in FinFET nodes require an accurate estimative, besides new design methodologies can for reducing the effects caused by them. According to the literature review, few works propose techniques to attenuate the impact of the process variations and soft errors, specifically for FinFET technologies. Moreover, there is a lack of circuit-level mitigation approaches exploring design changes to achieve more robust solutions. In this way, the information provided in this book advances the state-of-the-art providing:

- The evaluation of FinFET logic cells under process variability and radiation effects using a 7-nm FinFET predictive process design kit;
- The design of logic cells using four different circuit-level approaches to mitigate the impact caused by work-function fluctuations and soft errors;
- A trade-off between the gains and penalties of each approach regarding the area, performance, power consumption, SET pulse width, and SET cross section;
- The mitigation tendency of the circuit-level techniques when different levels of the process variation, transistor sizing, and LET were applied in the design.
- The circuit-level mitigation approaches explored in this book were the transistor reordering, and the insertion of decoupling cells, Schmitt Triggers, and sleep transistors.

© The Author(s), under exclusive license to Springer Nature Switzerland AG 2021
A. Zimpeck et al., *Mitigating Process Variability and Soft Errors at Circuit-Level for FinFETs*, https://doi.org/10.1007/978-3-030-68368-9_9

In general, all these techniques reduce the process variability effects and the soft error susceptibility, introducing fewer penalties, implementation cost, and design complexity when compared with the few alternatives available in the literature.

The transistor reordering technique can increase up to 8% the robustness of complex cells under process variations (5% of deviation). However, this method is not favorable to soft error mitigation, increasing the susceptibility of complex cells up to 20% for low LETs. Among all techniques evaluated, the transistor reordering presented fewer power and performance overheads. Besides, it has no area penalties.

The adopting of decoupling cells shows interesting outcomes for power variability control under levels of variation above 4%. On the other hand, this technique is efficient for reducing delay variability independently of the levels of variation. The greater improvements in the delay variability can be seen for lower levels of variation (1–3%). The design with decoupling cells decreases the soft error susceptibility around 10% for a high LET (58 MeV cm^2 mg^{-1}). The gains with lower LETs can reach a maximum of 4%. This technique presented a large area overhead but a smaller impact on power and performance metrics.

The best approach to control the process variations is the connection of a Schmitt Trigger in the output of FinFET cells. This technique can improve the delay variability up to 50%, mainly for manufacturing processes with 2–4% of WF deviations. For the power variability, higher robustness was obtained with higher levels of variations (4–5%). Moreover, all logic cells investigated are free of faults, even at the near-threshold regime (0.3 V) and under the influence of a high LET (58 MeV cm^2 mg^{-1}). However, as this technique adds six extra transistors, it introduces higher penalties in area and power.

The insertion of a sleep transistor between the pull-down network and the ground rail is advantageous for power variability control. The improvements for some cells exceed 10% for higher levels of variation (4–5%). On the other hand, the efficiency of this method for delay variability reduction depends on how the transistors are arranged with the sleep transistor in the pull-down network. This technique is also free of faults, even at the near-threshold regime (0.3 V) and under the influence of a high LET (58 MeV cm^2 mg^{-1}). The layout with a sleep transistor introduces a small area overhead, but the impact in power and performance metrics is very significant.

A design with larger decoupling cells (four or five fins) is advantageous for process variability mitigation, with an acceptable increase in overheads. The Schmitt Trigger with a larger number of fins also improves the sensibility to WF fluctuations, but the technical drawbacks involved are unacceptable. On the other hand, the sensitivity to process variability does not change significantly, increasing the sizing of the sleep transistor.

The more appropriate technique depends on the target application and its requirements regarding area, power consumption, and performance. The Schmitt Trigger technique presented the best results for process variability mitigation and radiation hardness. However, this approach introduces critical penalties on area, delay, and power consumption. In this way, if a designer wants to improve the reliability of the circuit by introducing more acceptable penalties, a design with decoupling cells is more indicated to control the delay variability. On the other hand, to reduce the power variability and soft error susceptibility, the best option is to use the sleep transistor technique.

Finally, the details provided in this book are useful to help:

- the semiconductor industry to obtain parametric yield improvements avoiding the many stages of the redesign;
- the designers to introduce a mitigation technique at the layout level for a given application knowing all the pros and cons of adopting it;
- the aerospace industry such as ONERA, the French Aerospace Lab, to design more reliable systems for the next generation of nano-satellite constellations.

9.1 Relevant Open Research

There are several possibilities of new experiments and test scenarios that can be done from this book. First, it is possible to extend this research by exploring two more potential approaches for improving FinFET logic cells' robustness: multi-finger design [1] and dual-gate pitch [2]. Both techniques should be implemented at the layout level.

An in-depth study can be done to discover the best places to put these cells in a chain of gates, such that it is not practical to apply decoupling cells, Schmitt Triggers, or sleep transistors in all logic cells of integrated circuits. It is also crucial to better understand each technique to attenuate the effects of process variations and radiation-induced soft errors. Moreover, some ways to reduce the technique drawbacks (area, power, delay) imposed by the circuit-level mitigation approaches explored in this work need to be investigated.

Another interesting topic to research is the influences on the design of WF fluctuations and the soft errors together, considering the worst-case scenario of faults to introduce the process variations on circuits. Furthermore, the impact of the process variations also can be estimated using as a metric the power-delay-product (PDP), which offers an accurate trade-off between the results of power and delay variability. Finally, a cell library for digital designs can be developed focusing on reliability issues, such as process variability mitigation and radiation hardness.

References

1. Forero, F., Galliere, J., Renovell, M., Champac, V.: Analysis of short defects in FinFET based logic cells. In: 2017 18th IEEE Latin American Test Symposium (LATS), pp. 1–6 (2017)
2. Marella, S.K., Trivedi, A.R., Mukhopadhyay, S., Sapatnekar, S.S.: Optimization of FinFET-based circuits using a dual gate pitch technique. In: 2015 IEEE/ACM International Conference on Computer-Aided Design (ICCAD), pp. 758–763 (2015)

Index

© The Author(s), under exclusive license to Springer Nature Switzerland AG 2021
A. Zimpeck et al., *Mitigating Process Variability and Soft Errors at Circuit-Level for FinFETs*, https://doi.org/10.1007/978-3-030-68368-9

Printed in the United States
by Baker & Taylor Publisher Services